Solving Problems in Fluid Mechanics

(GATE XE)

The Authors

Dr. Kshirod Kumar Dash, Ph.D., is working as an Associate Professor in the Department of Food Processing Technology, Ghani Khan Choudhury Institute of Engineering & Technology (GKCIET), Malda, West Bengal, India. Previously, he worked as an Assistant Professor in the Department of Food Engineering and Technology, Tezpur University, Tezpur, Assam. Dr. Dash completed his B.Tech in Agricultural Engineering from College of Agricultural Engineering & Technology, OUAT, Bhubaneswar; M.Tech. in Dairy and Food Engineering from IIT Kharagpur; and Ph.D. in Food Process Engineering from IIT Kharagpur. Dr. Dash has obtained his Post-Doc from Ohio State University, Columbus, Ohio, USA. He has guided several postgraduate and doctoral students and published more than 50 publications in peer-reviewed international journals and book chapters in edited volumes. He is a life member of different professional bodies in India and a member of IFT, USA. The research interests of Dr. Dash include Food Process Engineering, Unit Operations in Food Processing, Fluid Mechanics, Heat and mass transfer in food processing.

Dr. Paragmoni Kalita is presently working as an Assistant Professor in the Department of Mechanical Engineering, Tezpur University. Prior to joining his service at Tezpur University, Dr. Kalita also had worked in National Engineering Industries Limited, Defence Research and Development Organization (DRDO), Numarligarh Refinery Limited (NRL), and Oil India Limited (OIL). Dr. Kalita completed his B.E. in Mechanical Engineering from erstwhile MNREC (presently MNNIT), Allahabad in the year 2001; M.Tech. in Mechanical Engineering from erstwhile BHU (presently IIT BHU), Varanasi in the year 2004; and Ph.D. in Mechanical Engineering from IIT Guwahati in the year 2017. He is the recipient of the BHU Medal for standing first in the M.Tech (Mechanical Engineering) Examination for 2004. His areas of expertise are Computational Fluid Dynamics and Numerical Heat Transfer. Dr. Kalita has presented his research papers in various reputed national and international conferences. His research works also got published in prestigious international journals, including the International Journal of Applied and Computational Mathematics, Computers & Fluids, Journal of Computational Physics, and International Journal of Numerical Methods in Heat & Fluid Flow.

Solving Problems in Fluid Mechanics

(GATE XE)

Authors
Kshirod Kumar Dash
Paragmoni Kalita

2022
Astral Cracker
A Division of
Astral International Pvt. Ltd.
New Delhi – 110 002

ISBN: 978-93-5461-356-2

Published by : **Astral Cracker**
A Division of
Astral International Pvt. Ltd.
– ISO 9001:2015 Certified Company –
4736/23, Ansari Road, Darya Ganj
New Delhi-110 002
Ph. 011-43549197, 23278134
E-mail: info@astralint.com
Website: www.astralint.com

Printed at : **Replika Press Pvt. Ltd.**

PREFACE

Solving Problems in Fluid Mechanics (GATE XE) is designed to assist students in comprehending complex concepts and practicing fluid mechanic problems. In this book GATE, XE questions from 2009 to 2021 are included in the form of solved problems. The book covers problems in Flow and Fluid Properties, Kinematics of Fluid Motion, Integral Analysis for a Control Volume, Differential Analysis, Dimensional analysis, Internal flows, Bernoulli's Equation, and its Applications, Potential Flows, and External Flows. This book can be read by students preparing for GATE Engineering Sciences (XE), GATE Food Technology, GATE Mechanical Engineering, and GATE Agricultural Engineering. Industry and academic experts from mechanical engineering, food engineering, and related fields will also be greatly benefited from this book. Engineering sciences XE section comprises the subjects such as Engineering Mathematics, Fluid mechanics, Material Science, Solid Mechanics, Thermodynamics, Polymer Science and Engineering, Food technology, and Atmospheric and Oceanic Sciences.

We express our thanks to all the students and faculties of the Department of Food Processing Technology, Ghani Khan Choudhury Institute of Engineering & Technology (GKCIET), Malda, West Bengal; Department of Food Engineering and Technology, Tezpur University; Department of Mechanical Engineering, Tezpur University for their continuous encouragement to publish this book. We do hope that this book will be a valuable resource for students preparing for the GATE examination and the knowledge presented in the book will strengthen the Fluid Mechanics concepts of the students.

Dr. Kshirod Kumar Dash

Dr. Paragmoni Kalita

CONTENTS

	Preface	*v*
1.	GATE 2021 Fluid Mechanics	1
2.	GATE 2020 Fluid Mechanics	17
3.	GATE 2019 Fluid Mechanics	31
4.	GATE 2018 Fluid Mechanics	46
5.	GATE 2017 Fluid Mechanics	62
6.	GATE 2016 Fluid Mechanics	80
7.	GATE 2015 Fluid Mechanics	97
8.	GATE 2014 Fluid Mechanics	115
9.	GATE 2013 Fluid Mechanics	135
10.	GATE 2012 Fluid Mechanics	150
11.	GATE 2011 Fluid Mechanics	164
11.	GATE 2010 Fluid Mechanics	179
12.	GATE 2009 Fluid Mechanics	197

GATE 2021
FLUID MECHANICS

Q.1 The general relationship between shear stress and the velocity gradient $\frac{du}{dy}$ for a fluid is given by , $t = k(\frac{du}{dy})^2$ where k is a constant with appropriate units. The fluid is Newtonian if

(A) n> 1 (B) n< 1
(C) n= 1 (D) n= 0

Solution to Q.1

For a Newtonian fluid, the shear stress τ is proportional to the velocity gradient , $\frac{du}{dy}$ i.e., one can write

$$\tau \infty \frac{du}{dy}$$

$$or, \tau = \mu \frac{du}{dy} \tag{i}$$

here is a constant called the coefficient of dynamic viscosity. Comparing Eq. (i) with the given Eq. $t = k(\frac{du}{dy})^2$ we get n = 1

Hence, the answer is (C).

Q.2 Which one of the following options is TRUE?

(a) Path lines and streak lines are the same in an unsteady flow, and streamlines are tangential to the local fluid velocity at a point.

(b) Streamlines are perpendicular to the local fluid velocity at a point, and streamlines and streak lines are the same in a steady flow.

(c) Path lines and streak lines are the same in an unsteady flow, and streamlines and streak lines are the same in a steady flow.

(d) Streamlines are tangential to the local fluid velocity at a point, and streamlines and streak lines are the same in a steady flow.

Solution to Q.2

Streamlines are the imaginary lines in a flow field such that the tangent drawn at any point on the line shows the direction of fluid velocity at that point. Further, for a steady flow, the streamlines and streak lines are identical.

Hence, the answer is (D).

Q.3 If P_{in}= 1.2 Pa and P_{out}= 1.0 Pa are the average pressures at inlet and outlet respectively for a fully-developed flow inside a channel having a height of 50 cm, then the absolute value of average shear stress (in Pa) acting on the walls of the channel of length 5 m is:

(a) 0.005 (b) 0.02

(c) 0.01 (d) 0.05

Solution to Q.3

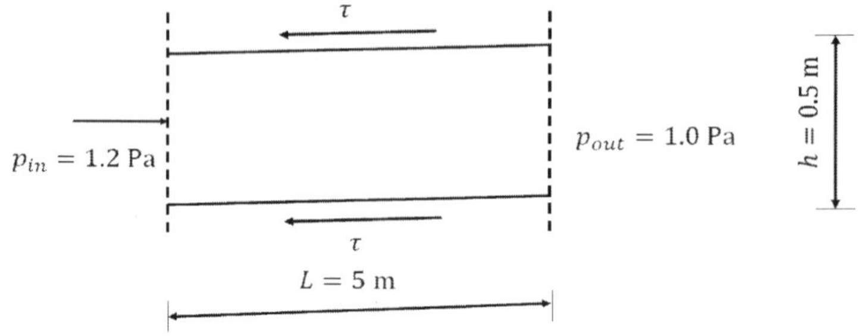

Fig. Q.3

With reference to Fig. Q.3, let the width of the channel perpendicular to the plane of the paper be 1 m.

Hence, the cross-sectional area for the flow is given by

$$A = h \times 1$$

$$or,\ A = h = 0.5 m^2 \qquad (i)$$

Let the average shear stress acting on the upper and lower wall of the channel be The total area on which the shear stress acts is given by

$$A\tau = 2 \times L \times 1$$

$$or\ \ A\tau = 2L = 10 m^2 \qquad (ii)$$

Now, since the flow is fully developed, hence the force balance gives

$$(P_{in} - P_{out}) \times A = \tau \times A_{\tau}$$

$$or\ \tau = (Pin - Pout) \times \frac{A}{A\tau}$$

$$= (1.2 - 1) \times \frac{0.5}{10} Pa$$

$$= 0.01\ Pa$$

Hence, the answer is (C).

Q.4 Consider the fully-developed flow of a Newtonian fluid (density ρ; viscosity μ) through a smooth pipe of diameter D and length L. The average velocity of the flow is V. If the length of the pipe is doubled, keeping V,D, ρ, μ constant, the friction factor.

(A) increases by two times
(B) remains the same
(C) decreases by two times
(D) increases by four times

Solution to Q.4

For a smooth pipe, the friction factor depends only on the Reynolds number $Re = \frac{\rho V D}{\mu}$. For the present case, ρ, V, D and μ are kept as constants. This means that also remains constant. Hence the friction factor also remains the same even thought the pipe length is doubled.

Hence, the answer is (B).

Q.5 The absolute value of pressure difference between the inside and outside of a spherical soap bubble of radius, R , and surface tension, γ is:

A. $\frac{2\gamma}{R}$
b. $\frac{\gamma}{R}$
C. $\frac{\gamma}{2R}$
d. $\frac{4\gamma}{R}$

Solution to Q.5

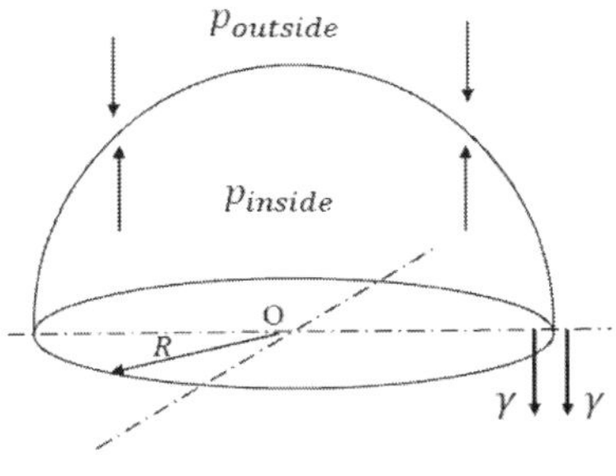

Fig. Q.5

With reference to Fig. Q.5, we consider one hemisphere of a soap bubble. It is evident that there are two surfaces – one inside and another outside of a soap bubble where the surface tension acts. Let the pressures inside and outside of the soap bubble be $P_{outside}$ and $P_{outside}$ respectively.

The projected area on which the force due to the pressure difference acts is equal to πR^2

The force exerted due to the pressure difference, $F_p = (P_{inside} - P_{outside})\pi R^2$

The force exerted due to the surface tension on two surfaces, $F_\gamma = 2 \times \gamma \times 2\pi R$

Therefore, the force balance gives

$$F_p = (P_{inside} - P_{outside})\pi R^2 = F_\gamma = 2 \times \gamma \times 2\pi R$$

$$\text{or, } P_{inside} - P_{outside} = \frac{4\gamma}{R}$$

Hence, the answer is (D).

Q.6 Which one of the following statements is TRUE about the continuity equation $\frac{\partial u}{\partial x}+\frac{\partial v}{\partial y}+\frac{\partial w}{\partial z}=0$ (where u,v, w are the velocity components alongthe x, y, and z coordinates respectively):

(A) The equation is valid only for steady incompressible flows.

(B) The equation is valid for both steady and unsteady incompressible flows.

(C) The equation is valid only for steady compressible flows.

(D) The equation is valid only for unsteady compressible flows.

Solution to Q.6

The general differential form of continuity equation in Cartesian coordinates is

$$\frac{\partial \rho}{\partial t}+\frac{\partial(\rho u)}{\partial x}+\frac{\partial(\rho v)}{\partial y}+\frac{\partial(\rho w)}{\partial z}=0$$

$$\text{or } \frac{\partial \rho}{\partial t}+u\frac{\partial \rho}{\partial x}+v\frac{\partial \rho}{\partial y}+w\frac{\partial \rho}{\partial z}+\rho\left(\frac{\partial u}{\partial x}+\frac{\partial v}{\partial y}+\frac{\partial w}{\partial z}\right)=0$$

$$\text{or } \frac{d\rho}{dt}+\rho\left(\frac{\partial u}{\partial x}+\frac{\partial v}{\partial y}+\frac{\partial w}{\partial z}\right)=0 \qquad \text{(i)}$$

In the case of an incompressible flow, $\rho = constant,\ i.e. \frac{d\rho}{dt}=0$ Since, $\rho \neq 0$ hence Eq. (i), for an incompressible flow becomes

$$\frac{\partial u}{\partial x}+\frac{\partial v}{\partial y}+\frac{\partial w}{\partial z}=0 \qquad \text{(ii)}$$

From Eqs. (i) and (ii), it is obvious that Eq. (ii) is valid for both steady and unsteady incompressible flows.

Hence, the answer is (B).

Q7. The head loss (K_L) associated with the flow entry of water to an internal passage depends on the shape of the entry. The following figure shows three different types of flow entry into a pipe. Which one of the following relationships correctly represents the head loss associated with the three different flow entries?

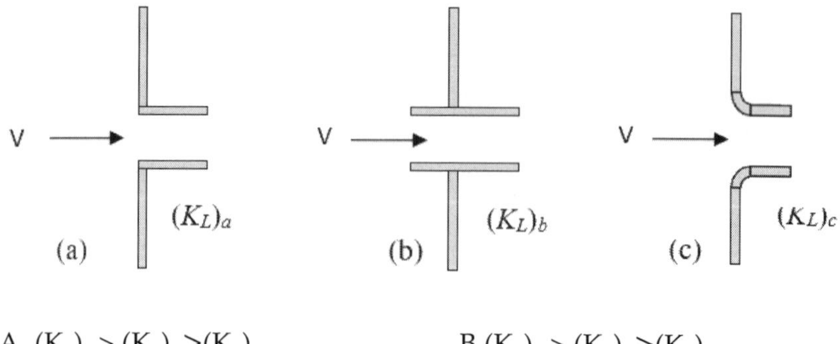

A. $(K_L)_a > (K_L)_b > (K_L)_c$ B $(K_L)_b > (K_L)_a > (K_L)_c$

C. $(K_L)_b \leq (K_L)_b = (K_L)_c$ D $(K_L)_b < (K_L)_a < (K_L)_c$

Solution to Q.7

The change in velocity of the fluid will be the highest in case (b), followed by case (a) and case (c), Therefore, the minor head loss will be the maximum in case (b) and minimum in case (c). Therefore, we can write. $(K_L)_b > (K_L)_a > (K_L)_c$.

Hence, the answer is (B).

Q.8 The form and friction drags together contribute to the total drag when flow of air occurs past any object. Two orientations of a finite flat plate are shown in the figure. In Orientation-1, the plate is placed perpendicular to the flow while in Orientation-2, the plate is placed parallel to the flow. If the velocity (V) of air in both orientations is the same, which one of the following options is TRUE?

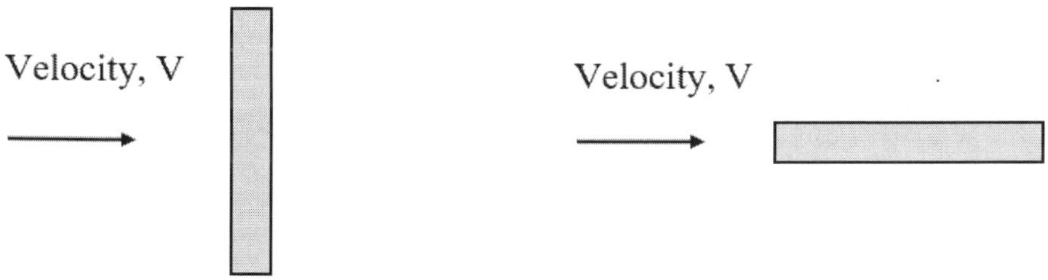

(Orientation-1) (Orientation-2)

(A) Orientation-1 has higher form drag and lower friction drag and Orientation-2 has lower form drag and higher friction drag

(B) Orientation-1 has lower form drag and lower friction drag and Orientation-2 has higher form drag and higher friction drag

(C) Orientation-1 has lower form drag and higher friction drag and Orientation-2 has higher form drag and lower friction drag

(D) Orientation-1 has higher form drag and higher friction drag and Orientation-2 has lower form drag and lower friction drag

Solution to Q.8

Orientation-1 offers higher area for the force due to pressure difference and lower area for the force due to skin friction compared with Orientation-2. Therefore, Orientation-1 has higher form drag and lower friction drag, whereas Orientation-2 has lower form drag and higher friction drag.

Hence, the answer is (A).

Q.9 A spherical ball is steadily supported against gravity by an upward air jet as shown in the figure. Take acceleration due to gravity to be $g = 10\ m/s^2$. The mass flow rate of air, reaching the ball, is 0.01 kg/s and the air reaches the ball at an upward velocity of 3 m/s. Neglecting the buoyancy force, and using the principle of integral momentum balance, the mass (in grams, up to one decimal place) of the ball is ________.

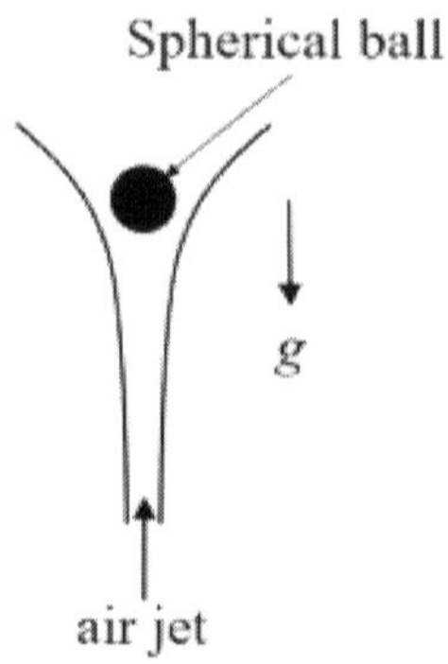

Solution to Q.9

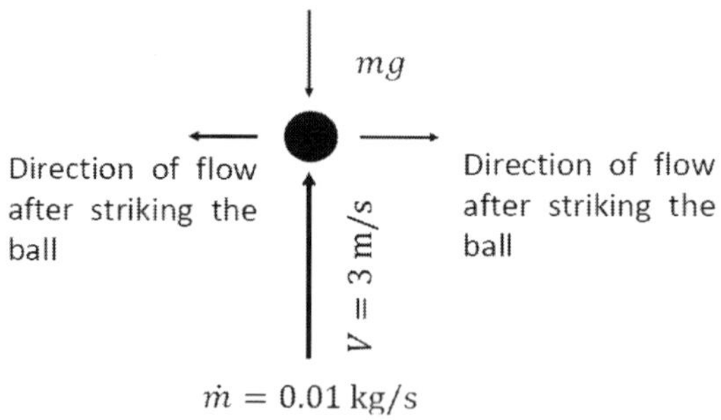

Fig. Q.9

With reference to Fig. Q.9 let us consider that the air turns by 90° after striking the ball w.r.t the initial direction of flow.

Hence the rate of change of momentum of air in the vertically upward direction

$$F_i = m\ (V - 0) = 0.01 \times 3\ N = 0.03N$$

The upward force due to change of momentum of air mut balance the weight of the ball, *i.e.*

$$\text{mg} = 0.03$$

$$\therefore m = \frac{0.03}{g} = \frac{0.03}{10} kg = 3\ gm$$

Hence, the answer is 3.

Q10. The incompressible flow of air over a curved surface having possible flow separation is schematically shown in the figure. Two zones P and Q are indicated in the figure. Which one of the following combinations is TRUE for zones P and Q?

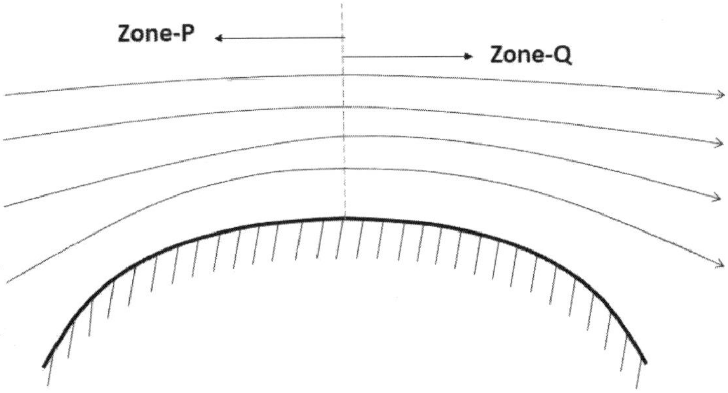

(a) Acceleration of flow, (b) Deceleration of flow, (c) Adverse pressure gradient, (d) Favorable pressure gradient, (e) No flow separation (f) Possible flow separation.

(A) P: (a), (d), (e) and Q: (b), (c), (f)

(B) P: (a), (d), (f) and Q: (b), (d), (f)

(C) P: (a), (c), (f) and Q: (a), (d), (e)

(D) P: (a), (c), (e) and Q: (a), (d), (f)

Solution to Q.10

In Zone-P, it can be shown using the potential flow theory that the flow accelerates, resulting in a decreasing pressure along the flow direction, *i.e.* a favorable pressure gradient. Naturally, there is no change of flow separation in Zone-P. The reverse phenomena happen in Zone-Q.

Hence, the answer is (A).

Q11. A spherical metal ball (of density ρ_s and diameter D), attached to a string, is exposed to a crossflow (of velocity $U\infty$) of a viscous fluid (of viscosity μ and density ρ_f). Due to the crossflow, the string makes an angle of inclination, θ, with the top surface as shown in the figure. The acceleration due to gravity is denoted by g. For this flow, Reynolds number, $Re = \frac{\rho f U\infty D}{\mu}$ << 1 and buoyancy force in the fluid is negligible compared toviscous force. Assuming the string to be weightless and offering negligible drag, the expression for θ is

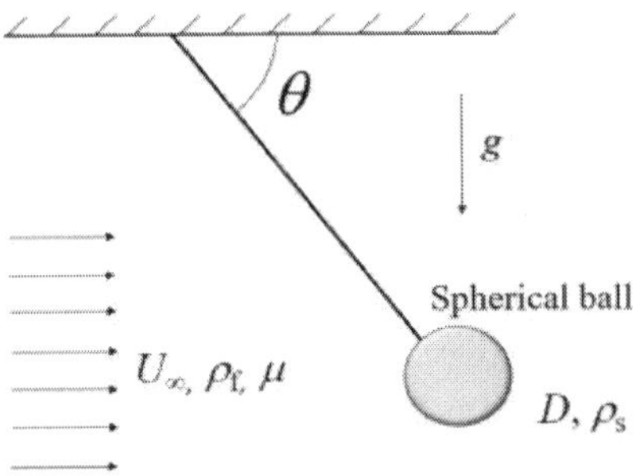

(A) $\tan^{-1}\left[\frac{1}{18}\frac{D^2\rho_s^2\, s\, g}{\mu U\infty \rho f}\right]$ (B) $\tan^{-1}\left[\frac{1}{18}\frac{D^2\rho\, fg}{\mu U\infty}\right]$

(C) $\sin^{-1}\left[\frac{2}{9}\frac{D^2\rho sg}{\mu U\infty}\right]$ (D) $\tan^{-1}\left[\frac{1}{18}\frac{D^2\rho sg}{\mu U\infty}\right]$

Solution to Q.11

With reference to Fig. Q.11, the Stokes' equation for the drag of a sphere gives the horizontal force acting on the sphere due to the drag induced by the flow

$$F_{vis} = 3\pi\mu D U_\infty \quad \text{(i)}$$

The vertically downward force acting on the sphere due to its own weight is

$$W = \frac{1}{6}\pi D^3 \rho sg \quad \text{(ii)}$$

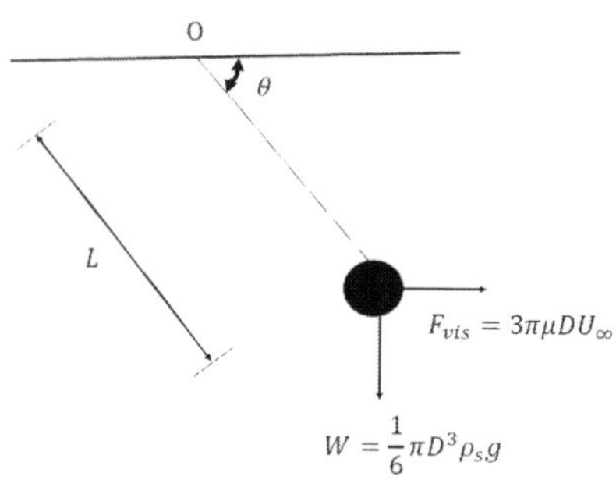

Fig. Q.11

At equilibrium moment due to the horizontal drag force must balance that due the weight of the ball, Let the length of the string be . Then balancing the moments due to the two forces about point O, we get

$$3\pi\mu DU\infty \times L\ \sin\theta = \frac{1}{6}\pi D^3\rho sg \times L\ \cos\theta$$

$$\therefore \theta = \tan^{-1}\left[\frac{1}{18}\frac{D^2\rho sg}{\mu U\infty}\right]$$

Hence, the answer is (D).

Q. 13 In a laminar, incompressible, fully-developed pipe flow of a Newtonian fluid, as shown in the figure, the velocity profile over a cross-section is given by $u = U\left(1-\frac{r^2}{R^2}\right)$ where U is a constant. The pipe length is L and the fluidviscosity is μ. The power P required to sustain the flow is expressed as $P = c\mu L U^2$ where c is a dimensionless constant. The value of the constant c (up to one decimal place) is ________.

Laminar, fully-developed pipe flow

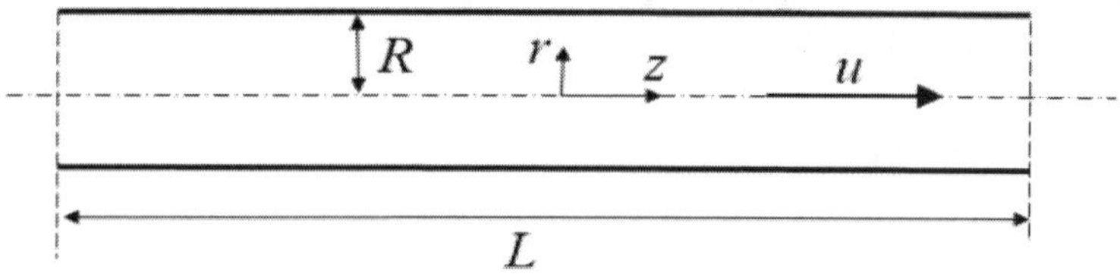

Solution to Q.13

If the average velocity through the pipe cross-section is Ū then

$$\overline{U} \times \pi R2 = \int_0^R U\left(1 - \frac{r^2}{R^2}\right) 2\pi r dr \quad \text{(i)}$$

After carrying out the necessary simplifications of Eq. (i) one gets]

$$\overline{U} = \frac{U}{2} \quad \text{(ii)}$$

The discharge through the pipe is given by

$$Q = \frac{\pi}{4} D^2 \overline{U} \quad \text{(iii)}$$

Let the density of the fluid be ρ. Then the head loss due to friction is given by

$$h_f = \frac{32\mu L\overline{U}}{\rho g D^2} \quad \text{(iv)}$$

The power required to sustain the flow against the head loss is given by

$$P = gQhf = \frac{32\mu L\overline{\overline{U}}}{D^2} \times \frac{\pi}{4} D^2 \,\overline{U}^2 = 2\pi\mu L U^2 \quad \text{(v)}$$

From Eq. (v) and the given expression for power requirement we can write

$$P = C\mu\, LU^2 = 2\pi\mu\, LU^2$$

$$\therefore\ C = 2\pi = 6.28$$

Hence, the answer is 6.28.

Q14 The two-dimensional velocity field V of a flow in a Cartesian coordinate system is given in dimensionless form by $V = (x^2 - axy)i + \left(bxy - \frac{y2}{2}\right)j$. Here, i and j are the unit vectors along the x and y directions respectively, a and b are independent of x, y and time. If the flow is incompressible, then the value of (a - b), up to one decimal place, is ____________ .

Solution to Q.14

The given velocity field is

$$V = (x^2 - axy)i + \left(bxy - \frac{y2}{2}\right)j \quad \text{(i)}$$

Hence, from Eq. (i) we have

$$u = x^2 - axy \; and \; v = bxy - \frac{y2}{2}$$

Since the flow is incompressible, therefore the continuity equation is

$$\frac{\partial u}{\partial x} + \frac{\partial v}{\partial y} = 0$$

or $$2 x- ay + bx-y =0 \qquad (ii)$$

Eq. (ii) has to be valid anywhere inside the flow domain. In order to get the value of (a−b) we must put x=1 and y = 1 in Eq. (ii). That gives us

$$2 \times 1- a \times 1 + b \times 1-1 =0$$

$$\therefore a -b =1$$

Hence, the answer is 1.

Q15 For the configuration shown in the figure, oil of density 800 kg/m³ lies-above water of density 1000 kg m³. Assuming hydrostatic conditions and acceleration due to gravity g = 10 m/s², the length L (in meters, up to one decimal place) of water in the inclined tube is .

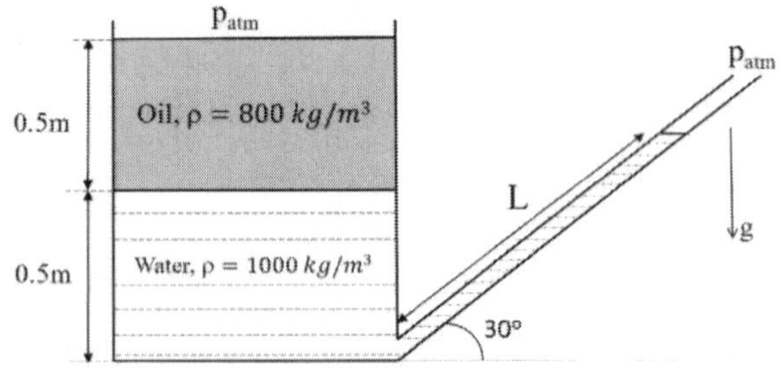

Solution to Q.15

For the given problem, we have

$$\rho_{oil} g h_{oil} + \rho_{water} g h_{waterl} = \rho_{water} g L sin\,(30°)$$

$$or, 800 \times 0.5 + 1000 \times 0.5 = 1000 \times L \times \frac{1}{2}$$

$$or, L= 1.8 \; m$$

Hence, the answer is 1.8.

Q16. A two-dimensional Eulerian velocity field is given (in m/s) by V=[(√5) x] i - [(√12)y]j, where x and y are the coordinates (in meters) in a Cartesian coordinate system. The magnitude of the acceleration (in m/s², up to one decimal place) of a fluid particle at x = 1 m and y = -1 m is _______ .

Solution to Q.16

The given velocity field is

$$\mathbf{V}=[(\sqrt{5})\ x]i - [(\sqrt{12})y]\mathbf{j} \quad \text{(i)}$$

Hence, from Eq. (i) we have

$$u = (\sqrt{5})\ x \text{ and } v = -\ (\sqrt{12})y$$

The acceleration along the x — direction is given by

$$ax = u\frac{\partial u}{\partial y} + v\frac{\partial u}{\partial y} = (\sqrt{5})\ x \times \sqrt{5} - (\sqrt{12})y \times 0 = 5x \quad \text{(ii)}$$

The acceleration along the y—direction is given by

$$ax = u\frac{\partial u}{\partial y} + v\frac{\partial u}{\partial y} = (\sqrt{5})\ x \times 0 - (\sqrt{12})y \times \sqrt{12} = -12\ y \quad \text{(ii)}$$

Therefore, the magnitude of acceleration at any point (x, y) is given by

$$a = \sqrt{a_x^2 + a_y^2} = \sqrt{25x^2 + 144y^2} \quad \text{(iii)}$$

From Eq. (iii) we get the acceleration at the point (x = 1, y = -1) as

$$a = \sqrt{25 \times (1)2 + 144 \times (-1)^2 = 13}$$

Hence, the answer is 13.

Q17. A large pump is to deliver oil at an average velocity (V) of 1.5 m/s. The pump has an impeller diameter (D) of 40 cm and the pressure rise across the pump is 400 kPa. To design this pump, a lab-scale model pump with an impeller diameter of 4 cm is to be used with water as the fluid. The viscosity (μ) of the oil is 100 times that of water, and the densities (ρ) of oil and water are identical. A complete geometric similarity is maintained between the model and prototype. If the pressure rise is a function only of V, D, and , the pressure rise (in kPa, up to one decimal place) acrossthe model pump is ________.

Solution to Q.17

Given for the prototype:

$$V_p = 1.5 \text{ m/s},\ D_p\ 0.4 \text{ m},\ \Delta p_p = 400 \text{ kpa}$$

Given for the model

$$D_m = 0.04\text{m}$$

Also given the following data

$$\mu_p = 100\mu_m,\ \rho_p = \rho_m$$

Now, for dynamic similarity

$$Re_m = Re_p$$

$$\frac{\rho_m V_m D_m}{\mu_m} = \frac{\rho_p V_p D_p}{\mu_p}$$

Now,

$$V_m = Vp\left(\frac{\rho_p}{\rho_m}\right)\left(\frac{D_p}{D_m}\right)\left(\frac{\mu_m}{\mu_p}\right)$$

$$= 1.5 \times (1) \times \left(\frac{0.4}{0.04}\right) \times \left(\frac{1}{100}\right) m/s$$

$$= 0.15 \text{m/s}$$

Now, $(C_p)_p = (C_p)_m$

$$or,\ \frac{(\Delta p)m}{\rho_m V_m^2} = \frac{(\Delta p)p}{\rho_p V_p^2}$$

$$or,\ (\Delta p)m\ \frac{\rho m\ V_m^2}{\rho p\ V_p^2}(\Delta p)\,p = 1 \times \left(\frac{0.15}{1.5}\right)^2 \times 400\ kPa = 4\ kPa$$

Hence, the answer is 4.

Q18. Water (density = 10^3 kg/m³) enters steadily into a horizontal pipe bend, which is part of a larger piping system, as shown in the figure. The volumetric flow rate of water is 0.1 m³/s. The gage pressure at the inlet is 500 kPa, while the exit is open to atmosphere. The x -component of the force on the support is F_x. The absolute value of F_x (in kN, up to one decimal place) is ________.

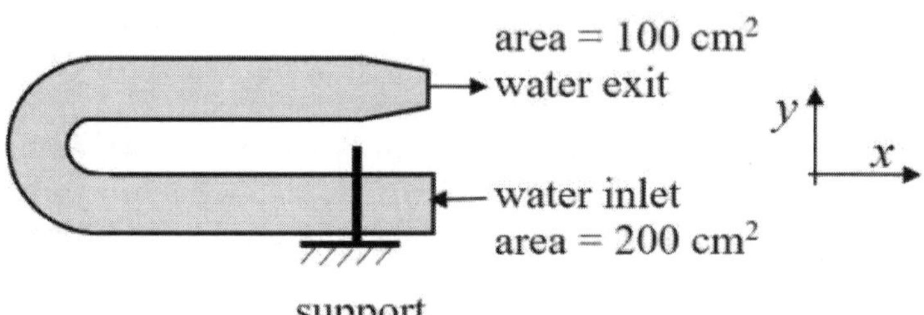

Solution to Q.18

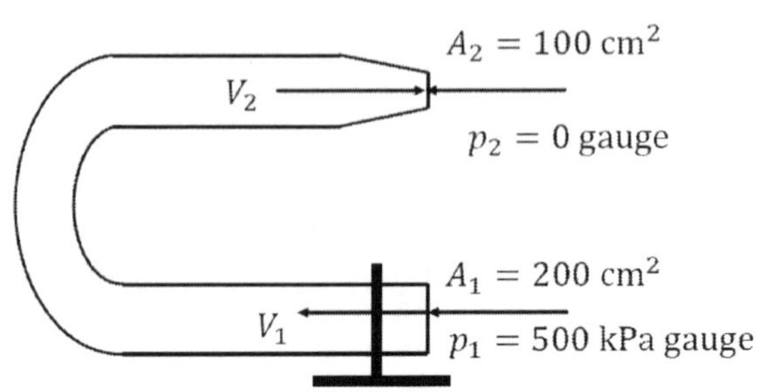

Fig. Q.18

Given data:

Discharge $Q = 01.m^3/s$

Density $\rho = 1000\ kg/m^3$

The rest of the data can be noted from Fig. Q.18.

The average velocity at inlet is obtained as

$$V_1 = \frac{Q}{A_1} = \frac{0.1}{200 \times 10^{-4}} \frac{m}{s} = 5m/s$$

The average velocity at outlet is obtained as

$$V_2 = \frac{Q}{A_2} = \frac{0.1}{100 \times 10^{-4}} \frac{m}{s} = 10m/s$$

Let the force exerted in the x- direction by the pipe bend on the fluid be F_x Therefore, the integral form of the momentum equation applied on the pipe gives

$$\rho Q (V_2 - V_1) = p_1 A_1 - p_2 A_2 + F_x$$

or

$$Fx = \rho Q (V_2 - V_1) + p_1 A_1 [\therefore P_2 = 0]$$

$$= 1000 \times 0.1 \times (10 - 5) + 500 \times 1000 \times 200 \times 10^{-4} N$$

$$\therefore F_x = 10.5 \text{ kN}$$

Hence, the answer is 10

Q19. Air (of density 0.5 kg/m^3) enters horizontally into a jet engine at a steady speed of 200 m/s through an inlet area of 1.0 m^2. Upon entering the engine, the air passes through the combustion chamber and the exhaust gas exits the jet engine horizontally at a constant speed of 700 m/s. The fuel mass flow rate added in the combustion chamber is negligible compared to the air mass flow rate. Also neglect the pressure difference between the inlet air and the exhaust gas. The absolute value of the horizontal force (in kN, up to one decimal place) on the jet engine is ________ .

Solution to Q.19

Given date

$$\rho_1 = 0.5\ kg/m^3$$

$$V_1 = 200\ m/s$$

$$A_1 = 1m^2$$

$$V_2 = 700\ m/s$$

The mass flow rate of air is obtained as

$$\dot{m} = \rho_1 A_1 V_1 = 0.5 \times 1 \times 200 \text{ kg/s} = 100 \text{ kg/s}$$

Neglecting the pressure difference across the engine, the absolute value of the force on the jet engine is given by

$$F = \dot{m}\,(V_2 - V_1) = 100 \times (700 - 200)\ \text{N} = 50\ \text{kN}$$

Hence, the answer is 50.

Q20. Water discharges from a cylindrical tank through an orifice, as shown in the figure. The flow is considered frictionless. Initially, the water level in the tank was $h_1 = 2$ m. The diameter of the tank is D = 1 m, while the diameter of the jet is d = 10 cm, and the acceleration due to gravity is $g = 10\ \text{m/s}^2$. The time taken (in seconds, up to one decimal place) for the water level in the tank to come down to $h_2 = 1$ m is__________.

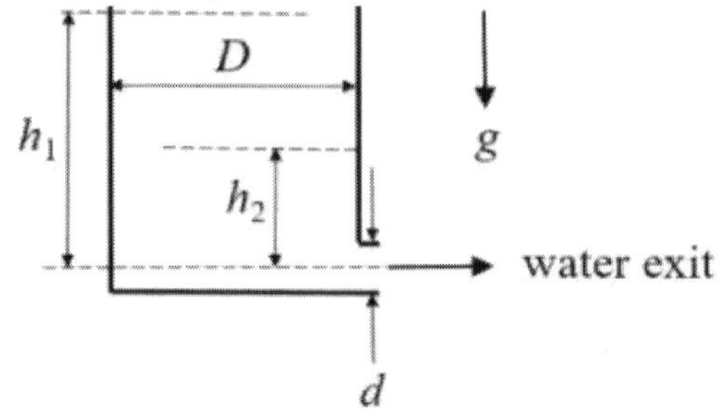

Solution to Q.20

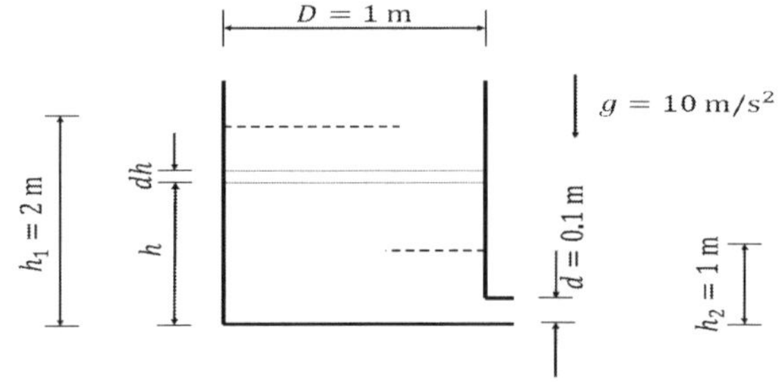

Fig. Q.20

Let at any instant of time the water level be located at a height of from the orifice level. Let the level decrease by an amount in an infinitesimally small time interval. Then the instantaneous discharge is given by

$$dQ = -\frac{\pi}{4}D^2\frac{dh}{dt} = \frac{\pi}{4}d^2\sqrt{5gh}$$

$$or\ dt = -\frac{D^2}{d^2}\frac{dh}{\sqrt{2gh}}$$

$$or,\ \int_0^t dt = -\frac{D^2}{d^2\sqrt{2g}}\int_{h1}^{h2}\frac{dh}{\sqrt{h}}$$

$$or,\ t = -\frac{D^2}{d^2\sqrt{2g}}[\,2\sqrt{h}\,]_{h1}^{h2}$$

$$= \frac{1^2}{(0.1)^2 \times \sqrt{2 \times 10}} \times 2 \times [\sqrt{2} - \sqrt{1}\,]s = 18.5s$$

Hence, the answer is 18.5.

Q21. Water discharges steadily from a large reservoir through a long pipeline, as shown in the figure. The Darcy friction factor in the pipe is 0.02. The pipe diameter is 20 cm and the discharge of water is 360 m^3/h. Water level in the reservoir is 10 m and acceleration due to gravity g = 10 m/s^2. If minor losses are negligible, the length L (in meters, up to one decimal place) of the pipeline is __________.

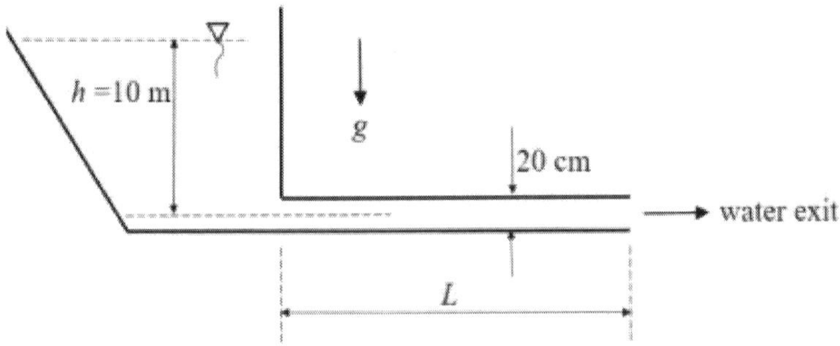

Solution to Q.21

Given data:

Head of water in the reservoir, $h = 10m$

Darcy friction factor, $f = 0.02$

Pipe diameter, $d = 20\ cm = 0.2m$

Discharge of water, $Q = 360m^3/h = 0.1m^3/s$

Acceleration due to gravity, $g = 10\ m^2/s$

Therefore, the average velocity of flow through the pipe is given by

$$\overline{V} = \frac{4Q}{\pi d^2} = \frac{4 \times 0.1}{\pi \times (0.2)^2} m/s = 3.18\ m/s$$

The limiting pipe length is obtained when the head loss due to friction equals the head of water in the reservoir, *i.e.*

$$h = \frac{fL\overline{V}^2}{2gd}$$

$$\therefore L = \frac{2dgh}{f\overline{V}^2} = \frac{2 \times 0.2 \times 10 \times 10}{0.02 \times (3.18)^2} m = 197.8m$$

Hence, the answer is 197.8.

Q.22 Water is flowing with a flow rate Q in a horizontal circular pipe. Due to the low pressure created at the venturi section (Section-1 in the figure), water from a reservoir is drawn upward using a connecting pipe as shown in the figure. Take acceleration due to gravity g = 10 m/s^2. The flow rate$Q$ = 0.1 m^3/s, D$_1$= 8 cm, and D$_2$= 20 cm. The maximum height (h, in meters, up to

one decimal place) of the venturi from the reservoir just sufficient to raise the liquid upto Section-1 is __________.

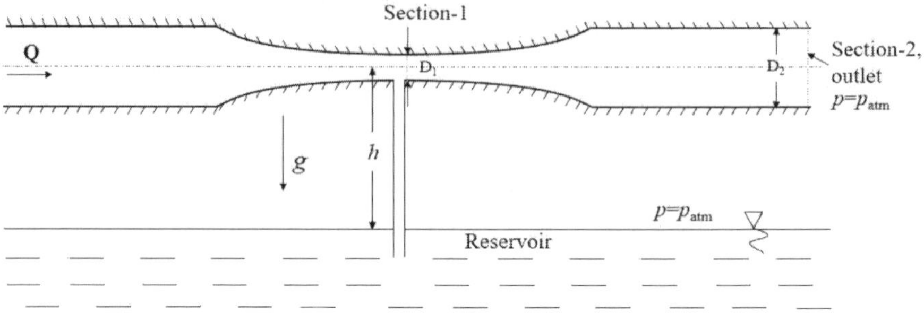

Solution to Q.22

Given data:

Discharge of water, Q = 0.1 m³/s

Diameter at the throat, D_1 = 8 cm = 0.08 m

Diameter at the exit, D_2 = 20 cm = 0.2 m

Pressure at the exit, p_2 = 1 atm

The velocity of water at the throat can be obtained as

$$V_1 = \frac{4Q}{\pi D_1^2} = \frac{4 \times 0.1}{\pi \times (0.08)^2} m/s = 19.89 m/s$$

The velocity of water at the exit can be obtained as

$$V_2 = \frac{4Q}{\pi D_2^2} = \frac{4 \times 0.1}{\pi \times (0.2)^2} m/s = 3.18 m/s$$

Let be the pressure at the venturi throat. Now applying Bernoulli's equation between the throat and the exit we get

$$p_1 + \frac{1}{2}\rho V_1^2 = p_2 + \frac{1}{2}\rho V_2^2$$

$$or,\ p_2 - p_1 = \frac{1}{2}\rho (V_1^2 - V_2^2)$$

$$= \frac{1}{2} \times 1000 \times \{(19.89)^2 - (3.18)^2\} N/m^2$$

$$= 192749.9 N/m^2$$

Since $p_2 = p_{atm}$ hence we can write

$$\rho g h = 192749.9$$

$$\therefore h = \frac{192749.9}{1000 \times 10} m = 19.2 m$$

Hence, the answer is 19.2.

GATE 2020
FLUID MECHANICS

Q. 1 Figures below show the velocity and shear stress profiles for the flow in a duct. In each option, '1' represents velocity profile and '2'represents shear stress profile.

(A) 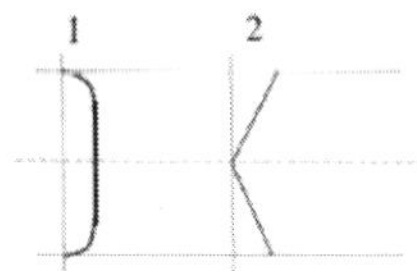

(B) 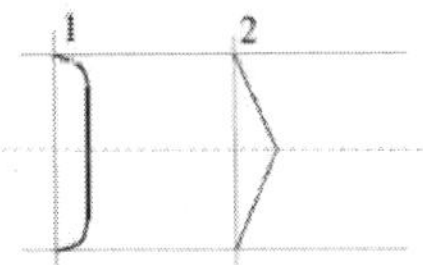

(C) 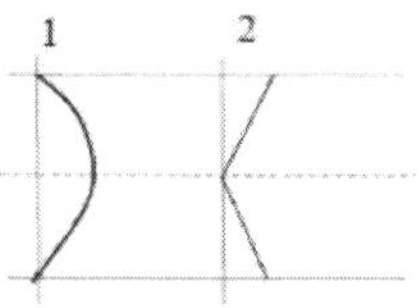

(D) 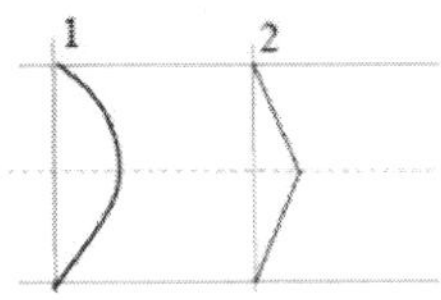

Choose the correct option that closely represents the turbulent flow condition.

Solution to Q.1

In a turbulent flow, the velocity profile possesses a sharp gradient near the solid wall and it is almost flat towards the centreline of the duct. Further, the velocity gradient along the radial direction is zero at the centreline. Therefore, the shear stress is maximum at the wall and is zero at the centreline. Therefore, the option (A) closely represents the turbulent flow condition.

Hence, the answer is (A).

Q.2 The variation of shear stress (τ) against strain rate (du/dy) is given in the Figure. Identify the line/curve among P, Q, R and S, that represents an ideal fluid.

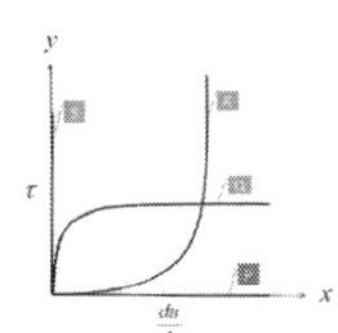

(A) S
(B) P
(C) Q
(D) R

Solution to Q.2

An ideal fluid possesses zero viscosity. This will lead to zero shear stress at all velocity gradients. Therefore, the curve P represents an ideal fluid.

Hence, the answer is (B).

Q.3 A body is under stable equilibrium in a homogeneous fluid, where CG and CB are center of gravity and center of buoyancy, respectively.

Two statements 'P' and 'Q' are given below:

P: For a fully submerged condition, CG should always be below CB

Q: For a floating body, CG need not be below CB

Choose the option that is valid for the present situation.

(A) P is False; Q is True when metacentre is below CG
(B) P is False; Q is True when metacentre is above CG
(C) P is True; Q is True when metacentre is below CG
(D) P is True; Q is True when metacentre is above CG

Solution to Q.3

For the stability of a fully-submerged body, CG should always be below CB in order to create a stabilizing couple in the event of a slight angular displacement from the equilibrium position.

For the stability of a fully-submerged body, the metacentric height should be positive in order to create a stabilizing couple in the event of a slight angular displacement from the equilibrium position.

Therefore, the answer is (D).

Q.4 A laminar hydrodynamic boundary layer over a smooth flat plate is shown in the Figure. The shear stress at the wall is denoted by (τ_w) Which one of the following conditions is correct.

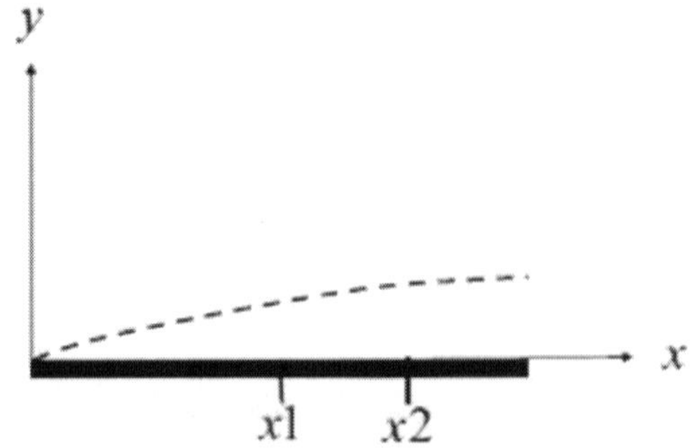

(A) Pressure is varying along 'x' and $(\tau_w)_{x1} > (\tau_w)_{x2}$
(B) Pressure is constant along 'x' and $(\tau_w)_{x2} > (\tau_w)_{x1}$
(C) Pressure is constant along 'x' and $(\tau_w)_{x1} > (\tau_w)_{x2}$
(D) Pressure is varying along 'x' and $(\tau_w)_{x2} > (\tau_w)_{x1}$

Solution to Q.4

For a hydrodynamic boundary layer over a flat plate, the pressure along the plate remains constant. Further, the boundary layer grows in its thickness (δ)

from zero at the leading edge till some finite value towards the trailing edge. Now, the velocity profile at the wall is higher at a location where the boundary layer is thinner compared with another location where the boundary layer is thicker. Since for the present problem $x_1<x_2$, hence $(\delta)_{x_1} < (\delta)_{x_2}$. Therefore, hence $(\tau_w)_{x_1} > (\tau_w)_{x_2}$.

Hence, the answer is (C).

Q.5 A non-dimensional number known as Weber number is used to characterize which one of the following flows,

(A) Motion of fluid in open channel

(B) Motion of fluid droplets

(C) Motion of fluid at high velocity

(D) Motion of fluid through a pipe

Solution to Q.5

Weber number is the ratio of surface tension force to the inertia force. These two forces are prominent in the case of motion of fluid droplets.

Hence, the answer is (B).

Q.6 A uniform approach flow is subjected to an unsteady and periodic flapping plate as shown in the Figure. Tracer is released to obtain flow visualization lines, which are marked as 'P', 'Q' and 'R'.

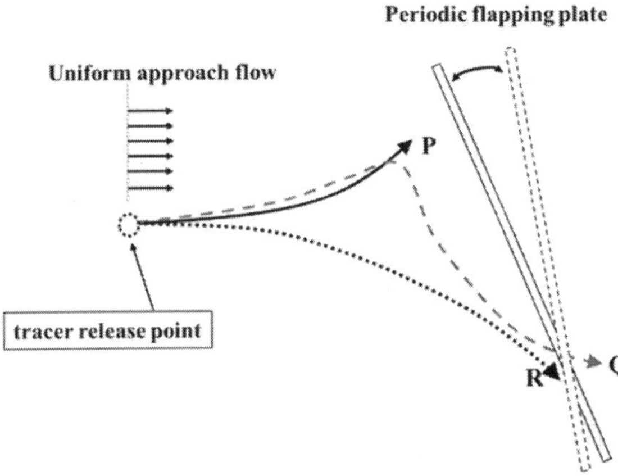

Choose; the correct option that the line 'R' represents

(A) Streakline

(B) Streamline

(C) Pathline

(D) Timeline

Solution to Q.6

A Streakline is the instantaneous picture of all the fluid particles that have passed through a certain point a some instant of time.

Hence, the answer is (A).

Q.7 The volume flow between any two points not lying on the same streamline in a flow field is equal to

(A) Change in strain rate between the points
(B) Change in vorticity between the points
(C) Change in potential function between the points
(D) Change in stream function between the points

Solution to Q.7

The change in stream function between two points not lying on the same streamline in a flow field is equal to the volume flow rate between the two points.

Hence, the answer is (D).

Q.No.8 A liquid flow through a horizontal smooth pipe of diameter 5 cm and discharges into a collection tank of dimension 50 cm x 50 cm x 50 cm. Time taken for a 10 cm rise of liquid level in the collection tank is 40 s.

The flow velocity in the pipe is ------------ m/s (rounded off to two decimal places).

Solution to Q.8

Volume flow rate, $Q = \frac{0.5 \times 0.5 \times 0.1}{40} \text{m}^3/\text{s} = 6.25 \times 10^{-4} \text{m}^3/\text{s}$

Hence, the flow velocity

$$\bar{V} = \frac{4Q}{\pi D^2} = \frac{4 \times 6.25 \times 10^{-4}}{\pi \times (0.05)^2} \text{m/s} = 0.32 \text{m/s}$$

Hence, the answer is 0.32.

Q. 9 The potential function for a two dimensional incompressible flow field is given as:

$$\phi = \frac{x^3}{3} - xy^2$$

Magnitude of the velocity vector at point (2,1) is ____________ m/s

Solution to Q.9

The x-velocity component, $u = \frac{\partial \varphi}{\partial x} = x^2 - y^2$

The y-velocity component, $v = \frac{\partial \varphi}{\partial y} = -2xy$

Hence, $u|_{(2,1)} = 3$ and $v|_{(2,1)} = -4$.

Therefore, the magnitude of velocity at the point $(2,1) = \sqrt{(3)^2 + (-4)^2} = 5$

Hence, the answer is 5.

Q.10 Column 1 represents a list of elementary plane flows and Column 11 represents flow past geometry obtained by superposition of these elementary plane flows.

Column I	Column II
P: Source, Sink, Uniform flow	1: Rankine half body
Q: Doublet, Uniform flow	2: Rotating Cylinder
R: Source, Uniform flow	3: Rankine oval
S: Doublet, Free vortex, Uniform flow	4: Cylinder

The correct match between Columns I and II is,

(A) P-3;Q-2;R-1;S-4
(B) P-1;Q-2;R-3;S-4
(C) P-3;Q-4;R-1;S-2
(D) P-1;Q-4;R-3;S-2

Solution to Q.10

Superposition of a Source, Sink and Uniform flow results in a Rankine oval, a combination of a Doublet and a Uniform flow results in a uniform flow past a cylinder, a combined flow of a Source and Uniform flow is equivalent to a Rankine half body and the superposition of a Doublet, Free vortex and Uniform flow is equivalent to uniform flow past a rotating cylinder.

Hence, the answer is (C).

Q.11 The velocity field for a flow is V = 5tî+ 2xzĵ + 2tyk̂, where / is time. Choose the correct option representing the total acceleration at (x,y,z,t)

(A) $5\hat{i} + 2(x+z)\hat{j} + 2(t+y)\hat{k}$
(B) $5\hat{i} + t(10z + 4xy)\hat{j} + (2y+4xzt)\hat{k}$
(C) $5\hat{i} + 2y\hat{k}$
(D) $2t(2xy + 5z)\hat{j} + 4xzt\hat{k}$

Solution to Q.11

The given velocity field is $\vec{V} = 5t\hat{i} + 2xz\hat{j} + 2ty\hat{k}.$. Hence, we have

$u = 5t, v = 2xz$ and $w = 2ty$

Therefore,

$$a_x = \frac{\partial u}{\partial t} + u\frac{\partial u}{\partial x} + v\frac{\partial u}{\partial y} + w\frac{\partial u}{\partial z}$$

$$= 5 + 5t \times 0 + 2xz \times 0 + 2ty \times 0$$

$$= 5$$

$$a_z = \frac{\partial w}{\partial t} + u\frac{\partial w}{\partial x} + v\frac{\partial w}{\partial y} + w\frac{\partial w}{\partial z}$$

$$= 2y + 5t \times 0 + 2xz \times 2t + 2ty + 0$$

$$= 2y + 4xyt$$

Therefore, the resultant acceleration is

$$\vec{a} = 5\hat{\imath} + t(10z + 4xy)\hat{\jmath} + (2y + 4xyt)\hat{k}$$

Hence, the answer is (B).

Q.12 An incompressible viscous fluid is placed between two infinite horizontal parallel plates as shown in Figure, The plates move in opposite direction with constant velocities U1 and U2.The pressure gradient in the x-direction is zero and the only body force is due to the fluid weight. The flow is steady, laminar and two-dimensional. Assume velocity component in 'y' direction to be zero.

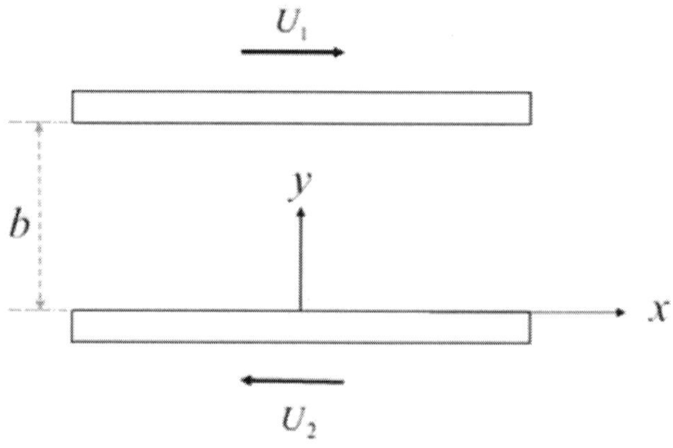

The correct expression for the velocity distribution between the plates is:

(A) $\left(\frac{U_1 + U_2}{b}\right)y - U_2$ (B) $\left(\frac{U_1 - U_2}{b}\right)y - U_1$

(C) $\left(\frac{U_1 + U_2}{b}\right)y + U_2$ (D) $\left(\frac{U_1 + U_2}{b}\right)y + U_1$

Solution to Q.12

The continuity equation is

$$\frac{\partial u}{\partial x} + \frac{\partial v}{\partial y} = 0$$

$$\frac{\partial u}{\partial x} = 0. \quad [\text{Since } v = 0] \qquad \ldots(i)$$

Now the x-momentum equation is

$$u\frac{\partial u}{\partial x}+v\frac{\partial u}{\partial y}=-\frac{1}{\rho}\frac{\partial p}{\partial x}+v\left(\frac{\partial^2 u}{\partial x}+v\frac{\partial^2 u}{\partial y^2}\right)$$

$$\frac{\partial^2 u}{\partial y^2}=0\left[\text{Since}\frac{\partial u}{\partial x}=\frac{\partial^2 u}{\partial x^2}=\frac{\partial p}{\partial x}=v=0\right]$$

$$\frac{\partial u}{\partial y}=C_1$$

$$u=C_1 y+C_{2.} \quad \text{...(ii)}$$

Now at y = 0, $u = -U_2$ and at $y = b$, $u = U_1$.

Therefore, from eq. (ii)

$$C_1\times 0+C_2=-U_2$$

$$C_2=-U_2. \quad \text{...(iii)}$$

And, $C_1 \times b - U_2 = U_1$

$$C_1=\frac{U_1+U_2}{b} \quad \text{...(iv)}$$

From eqs. (ii), (iii) and (iv) the velocity profile is obtained as

$$u=\frac{U_1+U_1}{b}y-U_1$$

Hence, the answer is (A).

Q.13 The stream function of a flow field is ψ=k(x2-y2x) where k is a constant. Which one of the following represents the vorticity?

(A) -2k (B) 2k (x+1)

(C) 2k (x-1) (D) -2k (x+1)

Solution to Q.13

Given: $\psi=k\left(x^2-y^2x\right)$

Hence,

$$u=\frac{\partial\psi}{\partial y}=-2kxy$$

$$v=-\frac{\partial\psi}{\partial x}=-2kx+ky^2$$

Since the flow is two dimensional, therefore the vorticity components along the x- and y-directions are zero, i.e. . $\Omega_x = \Omega_y = 0$. The vorticity component along the z-direction is

$$\Omega_z = \frac{\partial v}{\partial x} - \frac{\partial u}{\partial y}$$
$$= -2k + 2kx$$
$$= 2k(x-1)$$

Hence, the answer is (C).

Q.14 Consider a two dimensional, incompressible steady flow of a Newtonian fluid in which the velocity field is u = -2xy, v= y^2 -x^2. Pressure gradients in the x- and y- directions are :

(A) $\frac{\partial \rho}{\partial x} =- 2\rho\ (xy^2 + x^3), \frac{\partial \rho}{\partial y} =- 2\rho\ (x^2y\ + x^3)$

(B) $\frac{\partial \rho}{\partial x} =- 2\rho\ (xy^2 + x^3), \frac{\partial \rho}{\partial y} =- 2\rho\ (x^2y\ + y^3)$

(C) $\frac{\partial \rho}{\partial x} =- 2\rho\ (xy^2 + x^3), \frac{\partial \rho}{\partial y} =- 2\rho\ (x^2y\ - y^3)$

(D) $\frac{\partial \rho}{\partial x} =- 2\rho\ (xy^2 - x^3), \frac{\partial \rho}{\partial y} =- 2\rho\ (x^2y\ - y^3)$

Solution to Q.14

Given:

$u = -2xy$ and $v = y^2 - x^2$.

The flow is two dimensional, incompressible and steady, and the fluid is Newtonian.

The x-momentum equation is

$$u\frac{\partial u}{\partial x} + v\frac{\partial u}{\partial y} = -\frac{1}{\rho}\frac{\partial p}{\partial x} + v\left(\frac{\partial^2 u}{\partial x^2} + \frac{\partial^2 u}{\partial y^2}\right)$$

$$-2xy\times(-2y) + (y^2 - x^2)\times(-2x) = -\frac{1}{\rho}\frac{\partial p}{\partial x} + v\times(0+0)$$

$$-\frac{1}{\rho}\frac{\partial p}{\partial x} = 4xy^2 - 2xy^2 + 2x^3$$

$$\frac{\partial p}{\partial x} = -2\rho(xy^2 + x^3)$$

The y-momentum equation is

$$u\frac{\partial v}{\partial x}+v\frac{\partial v}{\partial y}=-\frac{1}{\rho}\frac{\partial p}{\partial y}+v\left(\frac{\partial^2 v}{\partial x^2}+\frac{\partial^2 v}{\partial y^2}\right)$$

$$-2xy\times(-2x)+(y^2-x^2)\times(2y)=-\frac{1}{\rho}\frac{\partial p}{\partial y}+v\times(-2+2)$$

$$-\frac{1}{\rho}\frac{\partial p}{\partial y}=4x^2y+2y^3-2x^2y$$

$$\boxed{\frac{\partial p}{\partial y}=-2\rho\left(x^2y+y^3\right)}$$

Hence, the answer is (A).

Q.15 A hydroelectric power plant takes in 30 m³/s of water through its turbine and discharges it to the atmosphere with V² = 2 m/s. The total head loss in the turbine and pen stock system is 20 m. (Assume turbulent flow with kinetic energy correction factor as 1.1. Density of water is 1000 kg/m3 and acceleration due to gravity, g is 10 m/s²).

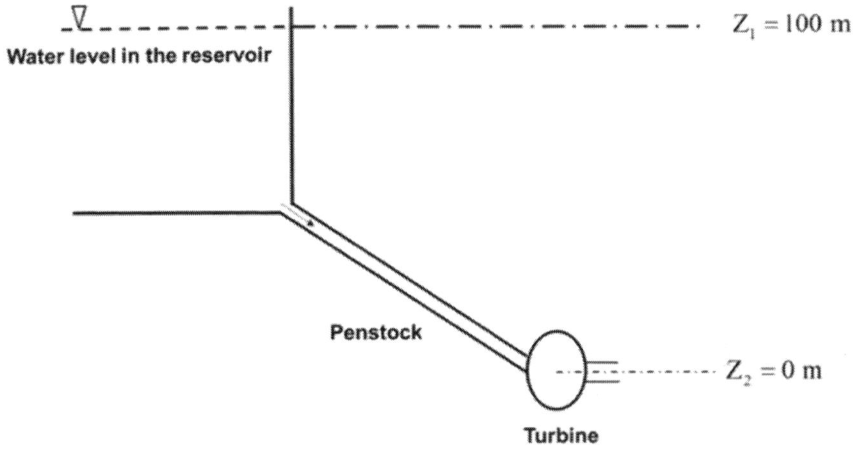

The net head available to the turbine for power generation is ------------------ m. (rounded off to one decimal place).

Solution to Q.15

The energy-balance equation between the free surface of water in the reservoir and the outlet of the turbine is

$$Z_1=\alpha\frac{V_2^2}{2g}+Z_2+H+h_f,\quad\left[\text{Since}\frac{p_1}{\rho g}=\frac{p_2}{\rho g}=0, V_1\approx 0\right] \qquad \ldots\text{(i)}$$

where α. H and h_f are the energy-correction factor, net head and head loss, respectively.

From eq. (i)

$$(C_D)_p = (C_D)_p \times \frac{(F_D)_m}{\frac{1}{2}r}$$

Hence, the answer is 79.78.

Q.16 Water flows at an average velocity, v of 10 m/s through a horizontal smooth tube of diameter, d 5 cm. The friction factor, f is 0.02. Head loss is obtained using Darcy - Weisbach relation $h_f = \frac{flv^2}{2gd}$ .The fluid pressure, p measured at various stations are reported in the table below. The length of the pipe, l between station 0 and station 6 is 6 m.

Station	0	1	2	3	4	5	6
p, kPa	304	273	255	240	226	213	-

If acceleration due to gravity, g = 10 m/s² and density of water =1000 kg/m³, then the fluid pressure at station 6 is __________ kPa (rounded off to one decimal place)

Solution to Q.16

$$hf = \frac{flv^2}{2gd}$$

$$= \frac{0.02 \times 6 \times 10^2}{2 \times 10 \times 0.05} \text{m}$$

$$= 12 \text{ m}$$

Hence, the pressure drop over the given length,

$$\Delta p = \rho g h_f = \frac{1000 \times 10 \times 12}{1000} \text{kPa} = 120 \text{ kPa}$$

Therefore, the pressure at station 6

$$= (304 - 120) \text{ kPa} = 184 \text{ kPa}$$

Hence, the answer is 184.

Q.17 A sphere model of 10 cm diameter is tested in water flowing at 2 m/s. The drag force is measured as 5 N. Prototype of 1.5 m in diameter is tested in air with dynamic similarity conditions. (Density of water is 1000 kg/m³, density of air is 1.2 kg/m³, viscosity of water is 0.001 Ns/m² and viscosity of air is 1.78 x 10-5 Ns/m²).

Drag force experienced by the prototype is ______N (rounded off to two decimal places).

Solution to Q.17

For dynamic similarity

$$Re^m = Re^p$$

$$\frac{\rho_m V_m D_m}{\mu_m} = \frac{\rho_p V_p D_p}{\mu_p}$$

$$V_p = V_m \left(\frac{\rho_m}{\rho_p}\right)\left(\frac{D_m}{D_p}\right)\left(\frac{\mu_p}{\mu_m}\right)$$

$$= 2 \times \left(\frac{1000}{1.2}\right) \times \left(\frac{0.1}{1.5}\right) \times \left(\frac{1.78 \times 10^{-5}}{0.001}\right) \text{m/s}$$

$$= 1.98 \, \text{m/s}$$

Now, $(C_D)_p = (C_D)_m$

$$(C_D)_p = \frac{(F_D)_m}{\frac{1}{2}\rho m\, V_m^2 \times \frac{\pi}{4} D_p^2} = \frac{5}{\frac{1}{2} \times 1000 \times 2^2 \times \frac{\pi}{4} \times (0.1)^2} = 0.3183$$

Therefore, the force experienced by the prototype

$$(F_D)_p = (C_D)_p \times \frac{1}{2}\rho_p V_p^2 \times \frac{\pi}{4} D_p^2$$

$$(F_D)_p = 0.0183 \times \frac{1}{2} \times 1.2 \times (1.98)^2 \times \frac{\pi}{4} \times (1.5)^2 \, N$$

$$= 1.32 \text{ N}$$

Hence, the answer is 1.32.

Q.18 A liquid of viscosity 1.74 x 10-3 Ns/m² is flowing through a horizontal capillary tube of diameter 0.5 mm. The flow in the tube is steady, incompressible, and fully developed laminar flow. The pressure drop across two locations spaced 1 m apart in the tube is 1.0 MPa. The flow rate in the tube ________________ mm³ /s.

Solution to Q.18

For a laminar flow through a duct,

$$\Delta p = \frac{32\mu LV}{D^2}$$

$$V = \frac{D^2 \Delta p}{D^2}$$

$$= \frac{(0.0005)^2 \times 1000000}{32 \times 1.74 \times 10^{-3} \times 1}$$

$$= 4.49 \text{m/s}$$

Therefore, the flow rate in the tube is

$$Q = V \times \frac{\pi}{4} D^2 = 4.49 \times 10^3 \times \frac{\pi}{4} \times (0.5)^2 \text{ mm}^3/\text{s} = 869.828 \text{ mm}^3/\text{s}$$

Hence, the answer is .

Q.19 A venturi meter with 75 mm diameter throat is placed in a 150 mm diameter pipeline carrying water at 25°C. The pressure drop between the upstream tap and the venturi throat is 40 kPa. (Density of water = 1000 kg/m^3).

The flow rate is ________________ m^3 /s (rounded off to three decimal places).

Solution to Q.19

The theoretical flow rate through the venturimeter is given by

$$Q = \frac{A_1 A_2}{\sqrt{A_2^1 - A_2^2}} \sqrt{\frac{2\Delta p}{\rho}} \tag{i}$$

Now,

$$A_1 = \frac{\pi}{4} D_1^2 = \frac{\pi}{4} \times (0.15)^2 \text{ m}^2 = 0.0177 \text{ m}^2 \tag{ii}$$

$$A_2 = \frac{\pi}{4} D_2^2 = \frac{\pi}{4} \times (0.075)^2 \text{ m}^2 = 0.0044 \text{ m}^2 \tag{iii}$$

Therefore, from eq. (i)

$$Q = \frac{0.0177 \times 0.0044}{\sqrt{(0.0177)^2 - (0.0044)^2}} \times \sqrt{\frac{2 \times 40000}{1000}} \text{ m}^2/\text{s} = 0.04 \text{ m}^3/\text{s}$$

Hence, the answer is 0.04.

Q.20 A water jet with velocity $\vec{V}_{jet}$ impinges normal to a moving flat plate with velocity $\vec{V}_{plate}$ such that the jet splits equally into two halves as shown in Figure.

The jet cross-sectional area is 2 cm^2, $\vec{V}_{jet}$ is 20 m/s and $\vec{V}_{plate}$ is 10 m/s and density of water is 1000 kg/m^3. Consider steady flow and neglect weight of the jet, weight of the plate and frictional losses.

The absolute value of the force required to keep the plate moving at constant velocity $\vec{V}_{plate}$ is ____________ N.

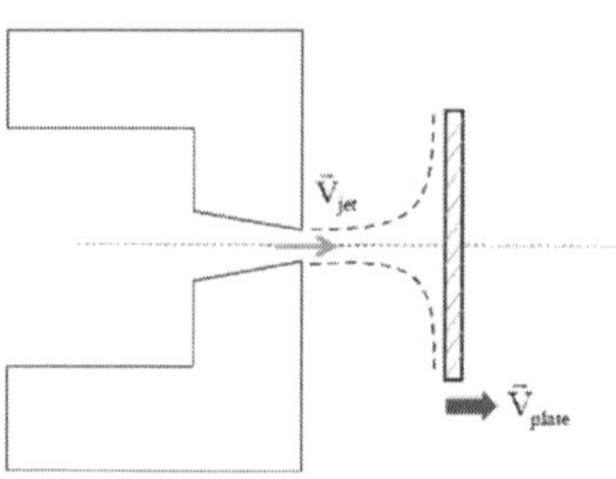

Solution to Q.20

The force exerted by the jet on the plate

$$F = \rho A_{jet}\left(V_{jet} - V_{plate}\right)\left\{\left(V_{jet} - V_{plate}\right) - 0\right\}$$

$$= \rho A_{jet}\left(V_{jet} - V_{plate}\right)^2$$

$$= 1000 \times 2 \times 10^{-4} \times (20 - 10)^2 \text{ N}$$

$$= 20 \text{ N}$$

Hence, the answer is 20.

Q.21 In an inverted manometer (as shown in the Figure), the pressure difference, $P_B - P_A$ is 100 kPa.

Use specific gravity of oil as 0.8, density of water as 1000 kg/m³, density of mercury as 13600 kg/m³ and acceleration due to gravity as 10 m/s². The height of the water column. H is ________________ cm. (rounded off to one decimal place).

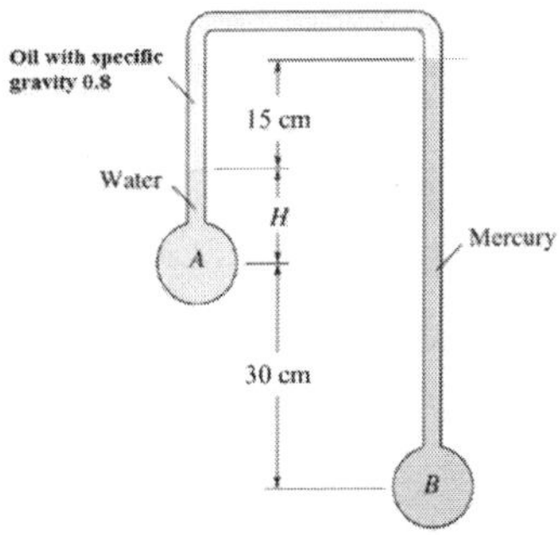

Solution to Q.21

The manometric equation is

$$p_A - \rho_w g H - \rho_o g \times 0.15 + \rho_m g \times (0.15 + H + 0.3) = p_B$$

$$H = \frac{p_B - p_A + \rho_o g \times 0.15 - \rho_m g \times 0.45}{g(\rho_m - \rho_w)}$$

$$= \frac{100000 + 800 \times 10 \times 0.15 - 13600 \times 10 \times 0.45}{10 \times (13600 - 1000)} \text{ m}$$

= 0.317 m =31.7 cm

Hence, the answer is 31.7.

Q.22 An incompressible, steady flow with uniform velocity condition at the inlet between parallel plates is shown in Figure. The flow develops into a parabolic laminar profile with $u=ay\,(y_0 - y)$at the downstream end. where 'a' is a constant. Assume unit depth of the plate. For $U_0 = 7.5$ cm/s, $y_0 = 3$ cm and the fluid with density, $\rho = 800$ kg/m^3

Solution to Q.22

The discharge per unit depth of the plates for a steady flow

$$Q = \int_{y=0}^{y=y_0} u dy$$

$$\int_{y=0}^{y=y_0} ay(y_0 - y)dy = U_0 y_0$$

$$\left[ay_0 \frac{y^2}{2} - a\frac{y^3}{3}\right]_{y=0}^{y=y_0} = U_0 y_0$$

$$a\frac{y_0^3}{6} = U_0 y_0$$

$$a = \frac{6U_0}{y_0^2}$$

$$= \frac{6 \times 7.5}{3^2}$$

$$= 5$$

Hence, the answer is 5.

GATE 2019
FLUID MECHANICS

Q.1 For stable equilibrium of a floating body, which one of the following statements is correct?

(A) Centre of gravity must be located below the centre of buoyancy.
(B) Centre of buoyancy must be located below the centre of gravity.
(C) Metacentre must be located below the centre of gravity.
(D) Centre of gravity must be located below the metacentre.

Solution to Q.1

For the stability of a floating body the metacentric height should be positive. This will happen only if the metacentre is above the centre of gravity.

Hence, the answer is (D).

Q.2 If u and v are the velocity components in the x- and y- directions respectively, the z- component of vorticity ω_z at a point in a flow field is:

(A) $\frac{\partial v}{\partial x} + \frac{\partial x}{\partial y}$ (b) $\frac{\partial v}{\partial x} - \frac{\partial u}{\partial y}$

(c) $\frac{\partial v}{\partial x} + \frac{\partial u}{\partial y}$ (d) $\frac{\partial v}{\partial y} - \frac{\partial u}{\partial x}$

Solution to Q.2

The vorticity in the x-y plane, i.e. about the z-axis is given by

$$\omega_2 = \frac{\partial v}{\partial x} - \frac{\partial u}{\partial y}$$

Hence, the answer is (B).

Q.3 In which one of the following devices the difference between static and total pressure is used to determine the flow velocity?

(A) Piezometer (B) Pitot static tube
(C) Orificemeter (D) Venturimeter

Solution to Q.3

A pitot static tube gives the flow velocity by measuring the difference between stagnation and static pressures.

Hence, the answer is (B).

Q.4 A golf ball is dimpled to make the flow turbulent and consequently to reduce the drag.

Turbulent flow reduces the drag on the golf ball because

(A) skin friction coefficient is lower in a turbulent flow.

(B) skin friction coefficient is higher in a turbulent flow.

(C) turbulent flow has a lower tendency to separate.

(D) turbulent flow has a higher tendency to separate.

Solution to Q.4

A golf ball is designed to experience as low drag as possible. It is known that the total drag on an object is because of skin friction drag and form drag. The skin friction drag is higher in the case of a turbulent flow than a laminar flow. However, a turbulent flow is less susceptible to flow separation than a laminar flow. This results in considerable reduction of the wake behind a cylinder, which in turn greatly reduces the form drag. The dimpled surface on a golf ball results in turbulent flow, which produces the overall effect of reduced total drag due to considerable reduction in form drag caused by a delayed separation.

Hence, the answer is (C).

Q.5 For a steady laminar incompressible boundary layer flow over a sharp-edged flat plate at zero incidence,

(A) The edge of the boundary layer is a streamline.

(B) The edge of the boundary layer is a path-line.

(C) The skin friction coefficient decreases as the distance from the leading edge increases.

(D) The skin friction coefficient remains constant all along the plate.

Solution to Q.5

For a steady laminar incompressible boundary layer flow over a sharp-edged flat plate at zero incidence, the boundary layer thickness continuously grows from zero at the leading edge till some finite value at the trailing edge. The rate of boundary layer growth is the maximum at the leading edge, and it reduces towards the trailing edge. The skin friction coefficient is directly proportional to the wall shear stress that is given as

$$\tau_w = \mu \frac{du}{dy}\bigg]_{y=0'}$$

where μ is the coefficient of dynamic viscosity and $\left.\frac{du}{dy}\right]_{y=0}$ is the velocity gradient at the wall. Smaller is the boundary layer thickness higher is the value of $\left.\frac{du}{dy}\right]_{y=0}$ and therefore higher is the skin friction coefficient. Accordingly, the skin friction coefficient is the highest at the leading edge and its value reduces along the plate in flow direction.

Hence, the answer is (C).

Q.6 The power input P to a centrifugal pump is a function of the volume flow rate Q, impeller diameter D , rotational speed n, fluid density p, dynamic viscosity μ, and surface roughness E . To carry out a dimensional analysis using Buckingham's rr theorem, which one of the following sets can be taken as the set of repeating variables?

(A) Q, 'Ω, D (B) Q, ε, D

(C) ε, D, ρ (D) D, ρ, 'Ω

Solution to Q.6

$P = P(Q,D,\Omega,\rho,\mu,\epsilon)$

$\therefore f(P,Q,D,\Omega,\rho,\mu,\epsilon)=0$

Therefore, $m = 7$ and $n = 3$. This means that the system can be reduced to (7-3)=4 non-dimensional groups. For this the number of repeating variables that need to be chosen is $n = 3$. The other conditions for choosing the repeating variables are – (i) the dependent variable (in this case P) should not be used as a repeating variable, and (ii) the repeating variables should contain all the number of fundamental dimensions among them.

Among the choices provided, the variables D, Ω satisfy the aforementioned conditions.

Hence, the answer is (D).

Q.7 Consider the two-dimensional laminar flow of water (μ = 0.001 N.s/m²) between two infinitely long parallel plates 0.1 m apart as shown in the figure below. The velocity profile at any location is given by u (y) = 100 (0.ly - y²) mis where y is in m. The magnitude of shear stress (in N/m²' rounded off to 2 decimal places) acting on the bottom plate is................................

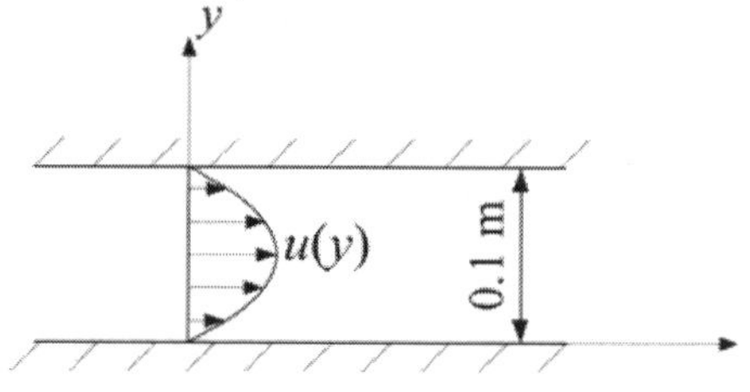

Solution to Q.7

The shear stress, acting on the bottom plate is given by

$$\tau_w = \mu \left.\frac{du}{dy}\right]_{y=0}$$

$$= \mu \left[\frac{d}{dy}\left\{100\left(0.1y - y^2\right)\right\}\right]_{y=0}$$

$$= 0.001 \times 10\,\text{N/m}^2$$

$$= 0.01\,\text{N/m}^2$$

Hence, the answer is 0.01 N/m^2.

Q.8 The maximum velocity in a fully developed laminar incompressible flow through a circular pipe of constant cross-sectional area is 6 mis. The average velocity (in mis) of the flow is................................

Solution to Q.8

For a fully developed laminar incompressible flow through a circular pipe of constant cross-sectional area, the average velocity $\overline{V}$ and the maximum velocity $V_{\max}$ are related as

$$\overline{V} = \frac{V_{\max}}{2}$$

Therefore for the present problem, $\overline{V} = \frac{6}{2}\,\text{m/s} = 3\,\text{m/s}$

Hence, the answer is 3.

Q.9 The theoretical discharge for the flow through an orifice meter is 40 m3/s. If the measured discharge in an experiment is 32 m^3/s, then the discharge coefficient (rounded off to one decimal place) is:..................

Solution to Q.9

Given:

Actual discharge, Q_{act} = 32 m^3/s

Theoretical discharge, Q_{theo} = 40 m^3/s

Therefore, the discharge coefficient is given by

$$C_d = \frac{Q_{act}}{Q_{theo}} = \frac{32}{40} = 0.8$$

Hence, the answer is 0.8.

Q.10 Consider the flow between two infinitely long parallel plates of large width separated by a distance 2*H*. The upper plate is moving with a constant velocity *U* while the lower plate is stationary. The volumetric flow rate per unit width of the plate is

(A)0.25 *UH* (B) 0.5 *UH*
(C) *UH* (D) 2 *UH*

Solution to Q.10

For a simple Couette flow the velocity profile is given by

$$u = U\left(\frac{y}{2H}\right)$$

Therefore, the volumetric flow rate per unit width of the plate can be obtained as

$$\begin{aligned} Q &= \int_{y=0}^{y=2H} u dy \\ &= \int_{y=0}^{y=2H} U\left(\frac{y}{2H}\right) dy \\ &= \frac{U}{2H} \times \left[\frac{y^2}{2}\right]_{y=0}^{y=2H} \\ &= \frac{U}{2H} \times \frac{4H^2}{2} \\ &= UH \end{aligned}$$

Hence, the answer is (C).

Q.11 The velocity field in Cartesian coordinates in a two-dimensional steady incompressible flow of a fluid with density is V = i j. Assuming no body and line forces, the magnitude of pressure gradient Δp at point (1, 1) is

(A) 2 ρ (B) ρ
(C) ρ/ 2 (D) ρ/2

Solution to Q.11

The given velocity field is $\vec{V} = x\hat{\imath} - y\hat{\jmath}$. Hence, we have
$u = x$ and $v = -y$.

Therefore,

$$\frac{\partial u}{\partial x} = 1 \qquad \text{(i)}$$

$$\frac{\partial u}{\partial y} = 0 \qquad \text{(ii)}$$

$$\frac{\partial^2 u}{\partial y^2} = 0 \quad \text{(iii)}$$

$$\frac{\partial^2 u}{\partial y^2} = 0 \quad \text{(iv)}$$

$$\frac{\partial v}{\partial x} = 0 \quad \text{(v)}$$

$$\frac{\partial v}{\partial x} = -1 \quad \text{(vi)}$$

$$\frac{\partial^2 v}{\partial x^2} = 0 \quad \text{(vii)}$$

$$\frac{\partial^2 v}{\partial y^2} = 0 \quad \text{(viii)}$$

Now the x-momentum equation for the given two-dimensional steady incompressible flow, is

$$u\frac{\partial u}{\partial x} + v\frac{\partial u}{\partial y} = -\frac{1}{\rho}\frac{\partial p}{\partial x} + \vartheta\left(\frac{\partial^2 u}{\partial x^2} + \frac{\partial^2 u}{\partial y^2}\right)$$

$$x \times 1 + (-y) \times 0 = -\frac{1}{\rho}\frac{\partial p}{\partial x} + \vartheta \times (0+0)$$

$$\frac{\partial p}{\partial x} = -\rho x \quad \text{(ix)}$$

Further, the y-momentum equation for the given two-dimensional steady incompressible flow, is

$$u\frac{\partial v}{\partial x} + v\frac{\partial v}{\partial y} = -\frac{1}{\rho}\frac{\partial p}{\partial y} + \vartheta\left(\frac{\partial^2 v}{\partial x^2} + \frac{\partial^2 v}{\partial y^2}\right)$$

$$x \times 0 + (-y) \times (-1) = -\frac{1}{\rho}\frac{\partial p}{\partial y} + \vartheta \times (0+0)$$

$$\frac{\partial p}{\partial y} = -\rho y \quad \text{(x)}$$

Hence, the magnitude of ∇p is given by

$$|\nabla p| = \sqrt{(-\rho x)^2 + (-\rho y)^2}$$

$$= \rho\sqrt{x^2 + y^2}$$

Therefore $|\nabla p|$ at point (1,1) will be

$$|\nabla p| = \rho\sqrt{1^2 + 1^2} = \sqrt{2}\rho$$

Hence, the answer is (A).

Q.12 A two-dimensional velocity field in Cartesian coordinates is defined by V= yi - xj, This flow is

(A) compressible and rotational (B) compressible and irrotational
(C) incompressible and rotational (D) incompressible and irrotational

Solution to Q.12

The given velocity field is $\vec{V} = y\hat{i} - x\hat{j}$. Hence, we have
$u = y$ and $v = -x$.

$$\text{Now, } \frac{\partial u}{\partial x} + \frac{\partial v}{\partial y} = 0 + 0 = 0.$$

Therefore, the continuity equation for an incompressible flow is satisfied.
Also, the angular velocity

$$\omega = \frac{1}{2}\left(\frac{\partial v}{\partial x} - \frac{\partial u}{\partial y}\right) = \frac{1}{2}(-1-1) = -1.$$

Therefore, the flow is rotational. Overall, the flow can be considered as incompressible and rotational.
Hence, the answer is (C).

Q.13 Assertion [a]: The streamlines in a free vortex flow are concentric circles.

Reasoning [r]: There exists only radial component for the velocity field in a free vortex flow.

(A) Both [a] and [r] are true and [r] is the correct reason for [a]
(B) Both [a] and [r] are true but [r] is not the correct reason for [a]
(C) [a] is true but [r] is false
(D) [a] is false but [r] is true

Solution to Q.13

The streamlines in a free vortex flow are concentric circles. However, the radial component of velocity in free-vortex flow is zero and there exists only the azimuthal-velocity component.
Hence, the answer is (C).

Q.14 The velocity components in Cartesian coordinates in a two-dimensional incompressible flow are u = e^y cos (x) and v=e^y sin (x). The magnitude of total acceleration at the point (1,1) is

(A) 0 (B) 1
(C) e (D) e^2

Solution to Q.14

Given:

$u = e^y \cos(x)$ and $v = e^y \sin(x)$.

∴ The acceleration along the x-direction is

$$\begin{aligned} a_x &= \frac{\partial u}{\partial t} + u\frac{\partial u}{\partial x} + v\frac{\partial u}{\partial y} \\ &= 0 + e^y \cos(x) \times e^y \{-\sin(x)\} + e^y \sin(x) \times e^y \cos(x) \\ &= 0 \end{aligned}$$

And the acceleration along the y-direction is

$$\begin{aligned} a_y &= \frac{\partial v}{\partial t} + u\frac{\partial v}{\partial x} + v\frac{\partial v}{\partial y} \\ &= 0 + e^y \cos(x) \times e^y \cos(x) + e^y \sin(x) \times e^y \sin(x) \\ &= e^2 y \times \{\cos^2(x) + \sin^2(x)\} \\ &= e^{2y} \end{aligned}$$

Therefore, the magnitude of acceleration at any point (x,y)

$$a = \sqrt{a_x^2 + a_y^2} = \sqrt{0 + e^{4y}} = e^{2y}.$$

The magnitude of acceleration at the point (1,1)

$$a|_{(1,1)} = e^{2\times 1} = e^2$$

Hence, the answer is (D).

Q.15 For steady laminar flow at zero incidence over a flat plate, the component of velocity parallel to the plate in the boundary layer is given by, u (y) = a + by + cy^2 where y is the distance measured normal to the flat plate. If is the coefficient of dynamic viscosity, is the velocity parallel to the wall at the edge of the boundary layer and is the boundary layer thickness, the wall shear stress is given by

(A) $\mu U/\delta$ (B) $2\mu U/\delta$
(C) $2\mu(U/\delta)^2$ (D) $3\mu U/\delta$

Solution to Q.15

Given:

The velocity profile is $u(y) = a + by + cy^2$

Now, for a boundary layer flow,

$$u(0) = 0$$

$$a + b \times 0 + c \times 0^2 = 0$$

$$a = 0 \tag{i}$$

Also, at the edge of the boundary layer $y = \delta$ and

$$u(\delta) = U$$

$$a + b \times \delta + c \times \delta^2 = U$$

$$b \times \delta + c \times \delta^2 = U \tag{ii}$$

The third boundary condition is

$$\left.\frac{du(y)}{dy}\right]_{y=\delta} = 0$$

$$b + 2c \times \delta = 0 \tag{iii}$$

Solving eqs. (ii) and (iii) we get,

$$b = 2\frac{U}{\delta} \text{ and } c = -\frac{U}{\delta^2}$$

and

Hence the velocity profile within the boundary layer becomes

$$u(y) = 2U\frac{y}{\delta} - U\left(\frac{y}{\delta}\right)^2$$

Hence, the wall shear stress is

$$\tau_w = \mu\left.\frac{du}{dy}\right]_{y=0}$$

$$= \mu \times \frac{2U}{\delta}$$

$$= \frac{2\mu U}{\delta} \tag{iv}$$

Hence, the answer is (B).

Q.16 A fluid with constant density of 1 kg/m3 flows past a semi-cylindrical structure with a freestream velocity of 2 mls as shown in the figure below. The difference in static pressure between points P and Q is 10 N/m2. If the gravitational acceleration g is 10 m/s2 and the flow is assumed to be potential, what is the radius (in m) of the semi-cylindrical structure?]

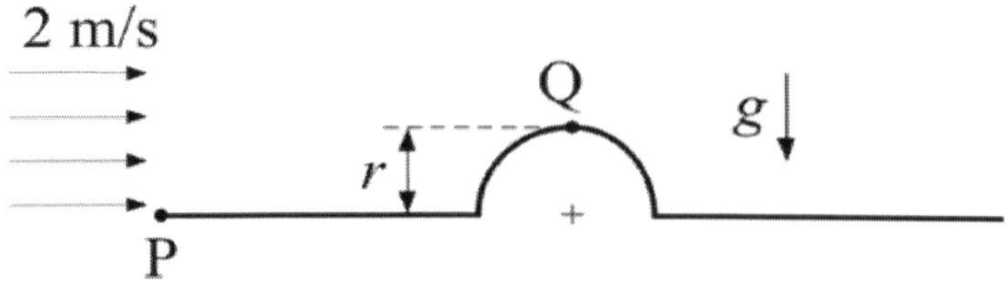

(A) 1 (B) 0.8
(C) 0.6 (D) 0.4

Solution to Q.16

Given:

Free-stream velocity, $U_0 = 2$ m/s

From potential flow theory, the velocity at point Q, $U = 2U_0 = 4$ m/s.

Let the free-stream pressure be p_0, the velocity at point Q be U and the pressure at point Q be p. For the potential flow we can apply Bernoulli's equation between points P and Q, which can be written as

$$\frac{Po}{\rho g} + \frac{U_0^2}{2g} = \frac{P}{\rho g} + \frac{U^2}{2g} + r$$

$$r = \frac{po - p}{\rho g} + \frac{U_0^2 - U^2}{2h}$$

$$= (\frac{10}{10} + \frac{2^2 - 4^2}{2 \times 10}) \text{m}$$

$$= 0.4 \text{ m}$$

Hence, the answer is (D).

Q.17 The mercury manometer shown in the figure below is connected to a water pipe at one end while the other end is open to the atmosphere. The density of water is 1000 kg/m3, the specific gravity of mercury is 13.6 and the gravitational acceleration g is 10 m/s². The gauge pressure (in kN/m2, rounded off to 2 decimal places) in the water pipe is ________

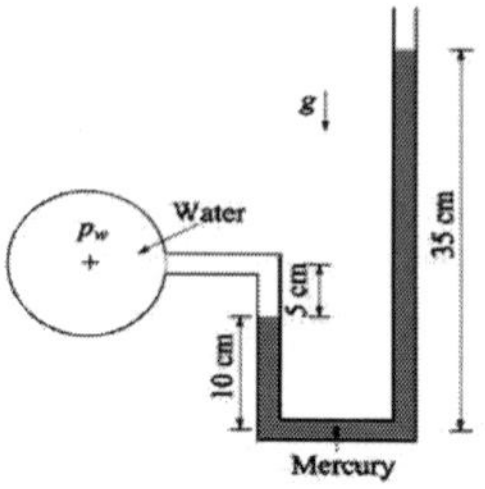

Solution to Q.17

Given:

The density of water, ϱ_w=1000 kg/m^3

The density of mercury, ϱ_m=13600 kg/m^3

Height of water column in the left limb of the manometer, h_w=5 cm=0.05 m

Mercury-level difference across the two limbs of the manometer, h_w=25 cm=0.25m

Let the atmospheric pressure be p_a. Hence for the given manometer we can write

$$p_w + \rho_w g h_w = p_a + \rho_m g h_m$$

$$p_w - p_a = \rho_m g h_m - \rho_w g h_w$$

$$p_{gauge} = \rho_m g h_m - \rho_w g h_w$$

$$= (13600 \times 10 \times 0.25 - 1000 \times 10 \times 0.05)\ \text{N/m}^2$$

$$= 33.5\ \text{kN/m}^2$$

Hence, the answer is 33.5 kN/m^2.

Q.18 Water (ρ= 1000 kg/m3,μ = 0.001 N.s/m^2) flows through a smooth circular pipe of radius 0.05 m. If the flow Reynolds number is 1000, then the pressure drop (in N/m^2, rounded off to 2 decimal places) over a length of 5 m will be ________

Solution to Q.18

Given:

The density, ρ =1000 kg/m^3

Coefficient of dynamic viscosity, μ = 0.001 N.s/m^2

Pipe radius, r = 0.05 m

Reynolds number, Re = 1000

Pipe length, L = 5 m

The pipe diameter, $D = 2\,r = 0.1$ m

Hence, the flow is laminar.

Now, if the flow velocity is V then

$$\text{Re} = \frac{\rho V D}{\mu}$$

$$V = \frac{\mu \times \text{Re}}{\rho D} = 0.01\,\text{m/s}$$

Therefore pressure drop over the pipe length is given by

$$\Delta p = \frac{32\mu LV}{D^2}$$
$$= \frac{32\times0.001\times5\times0.01}{(0.1)^2}\,\mathrm{N/m^2}$$
$$= 0.16\,\mathrm{N/m^2}$$

Hence, the answer is 0.16

Q.19 A uniform flow with a velocity of 2 mls in the -direction approaches a line source placed on the -axis at a distance of 0.1 m from the origin. If the origin is the stagnation point in the resulting flow, the strength of the source (in m²/s, rounded off to 2 decimal places) is ____________

Solution to Q.19

For the superposition of a uniform flow with a line source, the distance of the stagnation point is given by

$$r_s = \frac{q}{2\pi U} \tag{i}$$

where r_s is the distance of the stagnation point from the source, q is the discharge of the source per unit depth and is the uniform free-stream velocity. Hence the strength of the source is given by

$$\frac{q}{2\pi} = r_s U$$
$$= 0.1\times2\,\mathrm{m^2/s}$$
$$= 0.2\,\mathrm{m^2/s}$$

Hence, the answer is 0.2 m²/s

Q.20 In a steady incompressible flow of a fluid past a smooth stationary sphere, the drag force *F* depends on the flow velocity *U*, diameter *D*, and the dynamic viscosity μ and density ρ of the fluid. Experiments are conducted on the same sphere at the same flow velocity using two different fluids. The density of the second fluid is two times that of the first fluid. The dynamic viscosity of the second fluid is n times that of the first fluid. If the non-dimensional force F/ ($\rho U^2 D^2$) remains the same in both the experiments, the value of n is ____________

Solution to Q.20

Let the drag force *F* be related to the flow velocity *U*, diameter *D*, and the dynamic viscosity μ and density ρ as

$F = F(U,D,\mu,\rho)$

or, $f(F,U,D,\mu,\rho)=0$

From the above functional relationship, we can note that the number of variables $m = 5$ and the number of fundamental dimensions $n = 3$. Hence from Buckingham's π theorem, the functional relation can be expressed in terms of $m - n = 2$ non-dimensional numbers. Let the repeating variables be ρ, U and D.

Hence, the π-terms will be

$\pi_1 = \rho^{a1} U^{b1} D^{c1} \mu$

$\pi_2 = \rho^{a2} U^{b2} D^{c2} F$

By dimensional analysis it can be found that $\pi_1 = \frac{\rho UD}{\mu}$ and $\pi_1 = \frac{F}{\rho U^2 D^2}$ and

Now, since $\pi_2 = \frac{F}{\rho U^2 D^2}$ is constant for the present problem, hence $\pi_1 = \frac{\rho UD}{\mu}$ should also remain constant.

It is given that the experiments are carried out on the same cylinder and the flow velocity is also constant. Hence $U_1 = U_2 = U$ and $D_1 = D_2 = D$. . Also it is given that $\rho_2 = 2\rho_1$ and $\mu_2 = n\mu_1$

Since, $\pi_2 = \frac{F}{\rho U^2 D^2}$ is constant, hence

$$\frac{F_1}{\rho_1 U^2 D^2} = \frac{F_2}{\rho_2 U^2 D^2}$$

$$\frac{F_1}{\rho_1 U^2 D^2} = \frac{F_2}{2\rho_2 U^2 D^2}$$

$$F_2 = 2F_1$$

$$Also, \frac{\rho_1 UD}{\mu_1} = \frac{\rho_2 UD}{\mu_2}$$

$$\frac{\rho_1 UD}{\mu_1} = \frac{2\rho_1 UD}{n\mu_2}$$

$$n = 2$$

Hence, the answer is 2.

Q.21 An incompressible fluid flows past a flat plate as shown in the figure below with a uniform inlet velocity profile u =U and a parabolic exit velocity profile u = U(2n - n²), where u is the component of velocity parallel to the wall, y is the normal distance from the plate and n=y/δ . If the volume flow rate across the top surface of the control volume (CV) is $Q = pU\delta$ per unit width (perpendicular to the x - y plane) of the plate, the value of p (rounded off to 2 decimal places) is______

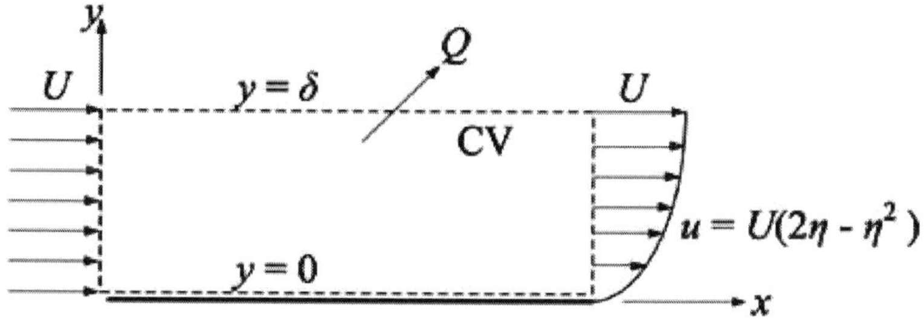

Solution to Q.21

Let the volumetric flow rate per unit width through the cross section at the leading edge be Q_1 and that be through the one at the trailing edge be Q_2.

Hence from the principle of mass conservation,

$$Q_1 = Q + Q_2$$

$$U\delta = pU\delta + \int_0^\delta u dy \tag{i}$$

$$\text{Now, } \int_0^\delta u dy$$

$$= \int_0^\delta U\left\{2\left(\frac{y}{\delta}\right) - \left(\frac{y}{\delta}\right)^2\right\} dy$$

$$= U\left[\frac{y^2}{\delta} - \frac{y^3}{3\delta^2}\right]_0^\delta$$

$$\int_0^\delta u dy = \frac{2}{3}U\delta \tag{ii}$$

Substituting eq. (ii) in eq. (i) we get

$$U\delta = pU\delta + \frac{2}{3}U\delta$$

$$p = 0.33$$

Hence, the answer is 0.33.

Q.22 A jet engine is to be tested on a thrust stand as shown in the figure below. The conditions prevailing in a typical test are as follows: Axial intake air velocity = 100 m/s; axial exhaust gas velocity = 250 m/s; intake cross-sectional area = 1 m^2; intake static pressure = 22 kPa (gauge); exhaust static pressure = 0 kPa (gauge); mass flow rate through the engine = 100 kg/s. The anchoring force (in kN) in axial direction on the thrust stand is ________

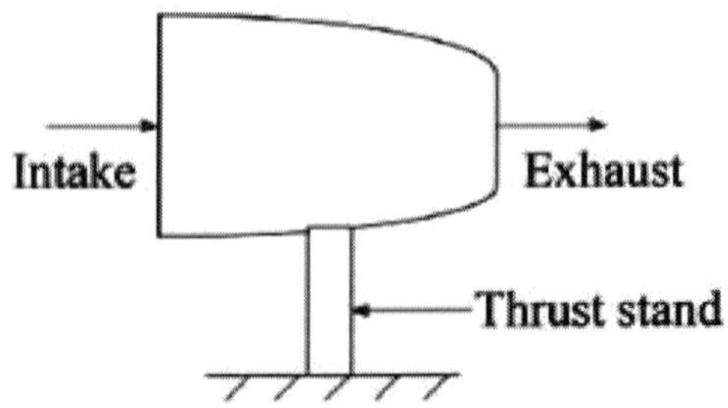

Solution to Q.22

The momentum equation across the engine is

$$\dot{m}(V_2 - V_1) = p_1 A_1 - p_2 A_2 + F$$

$$\begin{aligned} F &= \dot{m}(V_2 - V_1) - p_1 A_1 + p_2 A_2 \\ &= 100 \times (250 - 100) - (-22000) \times 1 + 0 \times 0.4 \text{ N} \\ &= 37000 \text{ N} \\ &= 37 \text{ kN} \end{aligned}$$

Hence, the answer is 37.

GATE 2018
FLUID MECHANICS

Q. 1 – Q. 9 carry one mark each & Q. 10 – Q. 22 carry two marks each.

Q1. Rheological diagram of different types of fluids is shown in figure. Column I represents the nature of the fluid and column II represents the curve showing the variation of shear stress against shear strain rate.

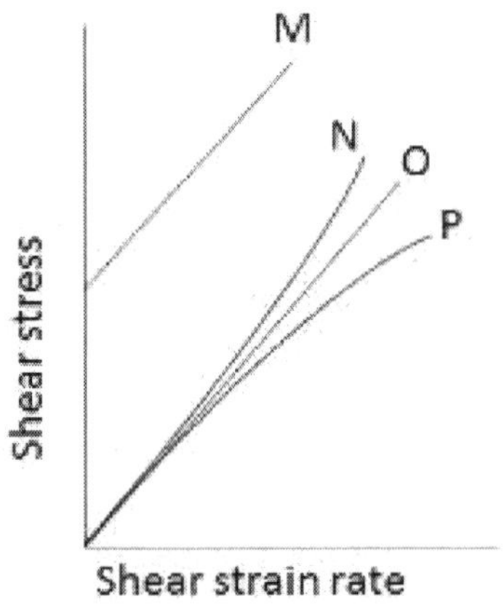

	Column I	Column II
(i)	Newtonian	M
(ii)	Shear thinning	N
(iii)	Shear thickening	O
(iv)	Bingham plastic	P

The most appropriate match between columns I and II is,

(A) (i) – O; (ii) – N; (iii) – P; (iv) – M
(B) (i) – O; (ii) – P; (iii) – N; (iv) – M
(C) (i) – P; (ii) – O; (iii) – M; (iv) – N
(D) (i) – P; (ii) – O; (iii) – N; (iv) – M

Solution:

i) **Newtonian:** For a Newtonian fluid, shear stress is directly proportional to shear strain rate, i.e., the viscosity (slope of the curve between shear stress vs shear strain rate) remains constant. Therefore, curve O represents a Newtonian fluid.

ii) Shear thinning: For a Shear thinning fluid, the apparent viscosity, i.e., the slope of the curve between shear stress strain rate decreases. Therefore, curve P represents a Shear thinning fluid.

iii) Shear thickening: For a Shear thickening fluid, the apparent viscosity, i.e., the slope of the curve between shear stress vs shear strain rate increases. Thus, curve N represents a Shear thickening fluid.

iv) Bingham plastic: for a Bingham plastic fluid, the viscosity remains constant but it requires some initial shear stress before deformation. Therefore, M represents a Bingham plastic liquid.

(i) O (ii) P
(iii) N (iv) M

Hence, the answer is (B).

Q2. In a two-dimensional, incompressible and irrotational flow, stream function $\psi = \psi(x, y)$ and velocity potential $\varphi = \varphi(x, y)$ exist. The velocities in x and y directions are non-zero. The product of $\left.\frac{dy}{dx}\right|_{\psi = \text{const}}$ and $\left.\frac{dy}{dx}\right|_{\varphi = \text{const}}$, is

(A) −1 (B) 0
(C) 1 (D) ∞

Solution:

In a two dimensional incompressible and irrational flow, the lines of constant stream function φ and constant velocity potential φ intersect each other orthogonally. Therefore, the product of their slope will be -1.

$$\text{i.e. } \left.\frac{dy}{dx}\right|_{\psi=const} \times \left.\frac{dy}{dx}\right|_{\psi=const} = -1$$

Hence, the answer is (A).

Q3. The inviscid flow past a rotating circular cylinder can be generated by the superposition of

(A) uniform flow, source and vortex
(B) uniform flow, doublet
(C) uniform flow, sink and vortex
(D) uniform flow, doublet and vortex

Solution:

The inviscid flow past a rotating circular cylinder can be generated by superposition of uniform flow, doublet and vortex flow.

Hence, the answer is (D).

Q4. The velocity field and the surface normal vector are given by, $\vec{V} = u\hat{i} + v\hat{j} + w\hat{k}$ and $\vec{n} = n_1\hat{i} + n_2\hat{j} + n_3\hat{k}$, respectively. If Euler equations are to be solved, the boundary condition that must be satisfied at the wall is,

(A) $\vec{V}.\vec{n} = 0$ (B) $\vec{V} = 0$

(C) $\nabla.\vec{V} = 0$ (D) $\vec{V} \times \vec{n} = 0$

Solution:

The velocity field is given by $\vec{V} = u\hat{i} + v\hat{j} + w\hat{k}$

The surface normal vector is given by, $\vec{n} = n_1\hat{i} + n_2\hat{j} + n_3\hat{k}$

Since the discharge across a wall is zero, hence the velocity vector and the surface normal vector are perpendicular to each other. Therefore, their dot product must be zero.

i.e. $\vec{V}.\vec{n} = 0$

Hence, the answer is (A).

Q5. The influence of Froude number is most significant in

(A) Capillary flows (B) Creeping flows

(C) Free surface flows (D) Compressible flows

Solution:

The influence of Froude number vector is most significant in free surface flow.

Hence, the answer is (C).

Q6. If the stream function $\psi(x, y)$ for a two-dimensional incompressible flow field is given as $2y(x^2 - y^2)$ the corresponding velocity field is

(A) $\vec{V} = 2(x^2 - 3y^2)\hat{i} + 4xy\,\hat{j}$ (B) $\vec{V} = 2(x^2 - 3y^2)\hat{i} - 4xy\,\hat{j}$

(C) $\vec{V} = 2(x^2 y)\hat{i} - 4xy\,\hat{j}$ (D) $\vec{V} = 2(x^2 y^2)\hat{i} + 4xy\,\hat{j}$

Solution:

Given:

Stream function is given by $\psi(x,y) = 2\,y(x^2-y^2)$

Now, we have, $u = \frac{\partial \psi}{\partial y}$ and $v = -\frac{\partial \psi}{\partial x}$

$\therefore u = \frac{\partial}{\partial y}\left(2yx^2 - 2y^3\right)$ and $v = -\frac{\partial}{\partial x}\left(2yx^2 - 2y^3\right)$

$u = 2x^2 - 6y^2$ and $v = -4xy$.

Therefore, the velocity field is given by $\vec{V} = (2x^2 - 6y^2)i - 4xy\hat{j} = 2(x^2 - 3y^2)\hat{\imath} - 4xy\hat{j}$.
Hence, the answer is (B).

Q7. Water is flowing in two different tubes of diameters *D* and 2*D*, with the same velocity. The ratio of laminar friction factors for the larger diameter tube to the smaller diameter tube is

(A) 0.5 (B) 1.0 (C) 2.0 (D) 4.0

Solution:

Given:

For tube 1, diameter, $d_1 = D$ and velocity, $V_1 = V$

For tube 2, diameter, $d_2 = 2D$ and velocity, $V_2 = V$.

Now, laminar friction factor is given by $f = \frac{64}{\text{Re}}$,

where the Reynolds number, Re is given by $\text{Re} = \frac{\rho V d}{\mu}$

with ρ = density of the fluid

d = diameter of the pipe

V = velocity of the flow

μ = coefficient of dynamic viscosity of the fluid

Hence, the friction factor f may be written as, $f = \frac{64\mu}{\rho V d} = \left(\frac{64\mu}{\rho}\right)\left(\frac{1}{Vd}\right) = \frac{K}{Vd}$

where $K = \frac{64\mu}{\rho}$ (constant)

Now, for tube 1, $f_1 = \frac{K}{d_1 V_1} = \frac{K}{D.V}$

and for tube 2, $f_2 = \frac{K}{d_2 V_2} = \frac{K}{2D.V}$

$$\therefore \frac{f_2}{f_2} = \frac{K}{2DV} \times \frac{D.V}{K} = \frac{1}{2} = 0.5$$

Hence, the answer is (A).

Q8. If the velocity field is $\vec{V} = xy^2 i + 4xy\hat{j}$ m/s, vorticity of the fluid element in the field at (x=1, y=2) in s^{-1} is ____.

Solution:

Given: Velocity field is given by $\vec{V} = xy^2 i + 4xy\hat{j}$ m/s

Now, vorticity is given by $\Omega = \left(\frac{\partial v}{\partial x}\right) - \left(\frac{\partial u}{\partial y}\right)$,

where $u = xy^2$ and $v = 4xy$.

Hence, $\Omega = 4y - 2xy$

$\therefore \Omega|_{x=1, y=2} = 4 \times 2 - 2 \times 1 \times 2 = 4$

Hence, answer is 4.

Q9. A pitot-static tube is used to measure air velocity in a duct by neglecting losses. The density of air is 1.2 kg/m³. If the difference between the total and static pressures is 1 kPa, the velocity of air at the measuring location, in m/s, is _______.

Solution:

Given:

The density of air, ρ =1.2 kg/m³.

The difference between total and static pressure, $p - p_s = 1 \text{ kPa} = 1 \times 10^3$ Pa.

Now, in a pilot tube the total pressure is given by

$$p = p_s + \frac{1}{2}\rho V^2$$

where p = Total pressure

p_s = Static pressure

$\frac{1}{2}\rho V^2$ = Dynamic pressure

$$\therefore \frac{1}{2}\rho V^2 = p - p_s$$

$$V = \sqrt{\frac{2(p - p_s)}{\rho}}$$

$$V = \sqrt{\frac{2\times10^3}{1.2}} \text{ m/s} = 40.82 \text{ ms}$$

Hence, the answer is 40.82

Q10. A parallelepiped of (2 m × 2 m) square cross-section and 10 m in length, is partially floating in water up to a depth of 1.2 m, with its longest side being horizontal. The specific gravity of the block is

(A) 0.8 (B) 0.6 (C) 0.5 (D) 0.4

Solution:

Given:

Dimensions of parallelepiped $2 \times 2 \times 10 = \text{m}^3$

Let, ρ_b = Density of the block

ρ_w = Density of water

Since the black is floating, therefore weight of the block must be equal to the buoyancy force.

i.e. $\rho_b vg = \rho_w v'g$

Where v= volume of the whole block

v'= volume of the portion of block which is submerged in water.

$\therefore \rho_b \times 2 \times 2 \times 10g = \rho_w \times 10 \times 2 \times 1.2$

$$\frac{\rho_b}{\rho_w} = \frac{10\times2\times1.2}{2\times2\times10} = 0.6$$

$\therefore$The specific gravity of the block is 0.6

Hence, the answer is (B).

Q11. The velocity field in a two-dimensional, unsteady flow is given by $\vec{V}(x,y,t) = 2xy^2\hat{\imath} + 3xyt\hat{\jmath}$ m/s m/s. The magnitude of acceleration of a fluid particle located at x = 1 m, y = 1 m at the time t = 1 s, in m/s², is

(A) 16.0 (B) 18.1 (C) 24.1 (D) 34.1

Solution:

Given:

Velocity field is given by, $\vec{V}(x,y,t) = 2xy^2\hat{\imath} + 3xyt\hat{\jmath}$ m/s

$$\therefore u = 2xy^2 \text{ and } v = 3xyt$$

Now, the x- component of acceleration is given by,

$$a_x = u\frac{\partial u}{\partial x} + v\frac{\partial u}{\partial y} + \frac{\partial u}{\partial t}$$

$$= 2xy^2 \times 2y^2 + 3xyt \times 4xy = 4xy^4 + 12x^2y^2t$$

$$\therefore a_x|_{x=1,y=1,t=1} = 4\times1\times1^4 + 12\times1^2\times1^2\times1 = 4+12 = 16\,\text{m/s}^2$$

And the y-component of acceleration is given by

$$a_y = u\frac{\partial v}{\partial x} + \frac{\partial v}{\partial y} + \frac{\partial v}{\partial t}$$

$$= 2xy^2 \times 3yt + 3xyt \times 3xt + 3xy$$

$$= 6xy^3t + 9x^2yt^2 + 3xy$$

$\therefore ay|_{x=1,y=1,t=1} = 6 \times 1 \times 1^3 \times 1 + 9 \times 1^2 \times 1 \times 1^2 + 3 \times 1 \times 1$

$=6+9+3=18$

∴The magnitude of acceleration is given by

$$a = \sqrt{(16)^2 + (18)^2} = 24.08 = 24.1 \text{m/s}^2$$

Hence, the answer is (C).

Q12. In a two-dimensional, incompressible and irrotational flow, fluid velocity (v) in the y-direction is given by $v = 2x - 5y$. The velocity (u) in the x-direction is

(A) $u = 2x + 5y$ (B) $u = 2x + 5y$

(C) $u = 5x + 2y$ (D) $u = 5x - 2y$

Solution:

Given:

The y-component of velocity, $v = 2x - 5y$.

For the flow to be incompressible, it must satisfy the continuity equation

$$\text{i.e.} \frac{\partial u}{\partial x} + \frac{\partial v}{\partial y} = 0$$

$$\frac{\partial u}{\partial x} + (0 - 5) = 0$$

$$\frac{\partial u}{\partial x} = 5$$

$$du = 5dx$$

$$du = 5x + f(y) \tag{i}$$

Now, for the flow to be irrational, we have

$$\text{i.e.} \frac{1}{2}\left(\frac{\partial v}{\partial x} - \frac{\partial u}{\partial y}\right) = 0$$

$$\left(\frac{\partial v}{\partial x} - \frac{\partial u}{\partial y}\right) = 0$$

$$2 - \frac{\partial u}{\partial y} = 0$$

$$\frac{\partial u}{\partial y} = 2$$

$$du = 2dy$$

$$u = 2y + f(x)$$

∴The x-component of velocity is given by

$$u = 5x + 2y$$

Hence, the answer is (C).

Q13. A two-dimensional laminar viscous liquid film of constant thickness (h) steadily flows down an incline as shown in figure. Acceleration due to gravity is g. If the velocity profile in the liquid film is given as, $u = ky(2h - y)$; $v = 0$, the value of constant k is

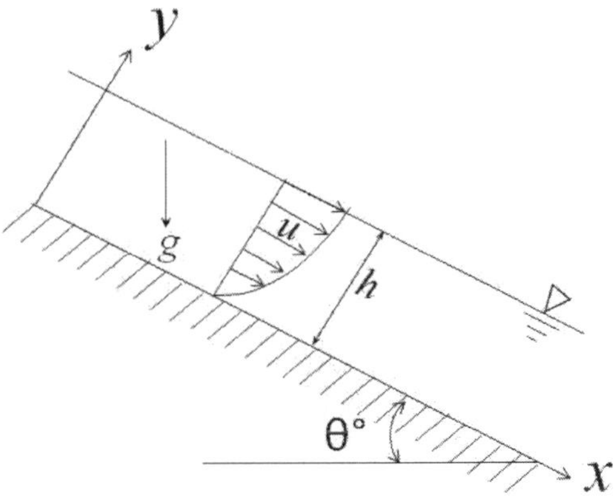

(A) $\dfrac{\rho g \sin\theta}{2\mu}$ (B) $\dfrac{\rho g \cos\theta}{2\mu}$

(C) $\rho g \sin\theta$ (D) $\rho g \cos\theta$

Solution:

Given:

Thickness viscous liquid film= h

Velocity profile in the liquid film, $u=ky(2h-y)$; $v=0$.

Now, writing the Navier-stokes equation in x-direction, we have

$$\rho\left[u\frac{\partial u}{\partial x}+v\frac{\partial u}{\partial y}+w\frac{\partial u}{\partial z}\right]=F_x-\frac{\partial p}{\partial x}+\mu\left(\frac{\partial^2 u}{\partial x^2}+\frac{\partial^2 u}{\partial y^2}+\frac{\partial^2 u}{\partial z^2}\right)$$

$$\rho[ky(2h-y)\times 0+0+0]=\rho g\sin\theta-\frac{\partial p}{\partial x}+\mu[0+k(0-2)+0]$$

$$0=\rho g\sin\theta-\frac{\partial P}{\partial x}-2\mu k$$

$$2\mu k=\rho g\sin\theta-\frac{\partial p}{\partial x}$$

$$2\mu k=\rho g\sin\theta-0$$

$$2\mu k=\rho g\sin\theta$$

$$k=\frac{\rho g\sin\theta}{2\mu}$$

Note: - F_x is the body force in x-direction, which is equal to $\rho g sin\theta$. Moreover, since the film thickness is constant, pressure at same level from the surface will be same, i.e., there is no pressure gradient along the x-direction. Therefore $\dfrac{\partial p}{\partial x}=0$.

Hence, the answer is (A).

Q14. A water jet of 100 mm diameter issuing out of a nozzle at a speed of 50 m/s strikes a vane and flows along it as shown in figure. The vane is attached to a cart which is moving at a constant speed of 20 m/s on a frictionless track. The jet is deflected at an angle of 30. Take the density of water as 1000 kg/m^3. Neglecting the friction between the vane and the fluid, the magnitude of the force exerted by water on the cart in the x-direction, in N, is ______

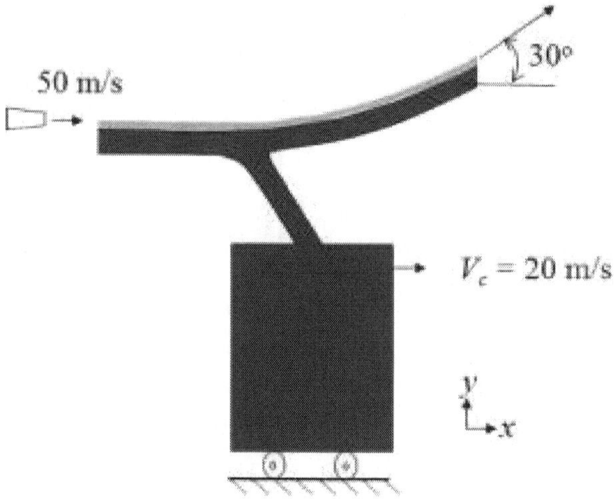

Solution:

Diameter of water jet, D=100 mm=0.1 m

Speed of the water jet, V= 50 m/s

Speed of the cart, V_c= 20 m/s

Angle of deflection, $\theta = 30°$

Density of water, ρ = 1000 kg/m^3

Now, the initial velocity of water jet relative to the cart, , V_1 = (50-20) m/s = 30 m/s (in the x-direction)

Since the flow is inviscid, hence the speed of the jet relative to the moving vane will remain the same, i.e. 30 m/s. However, at the outlet the jet is deflected by an angle 30°. Therefore, the final velocity of water jet in x- direction at the outlet

$V_2 = 30 \times \cos(30°)$ m/s = 25.98 m/s

Hence, the force exerted by water jet on cart in the x-direction is given by

$F_x = \rho Q(V_1 - V_2)$

$$= 1000 \times \frac{\pi \times (0.1)^2}{4} \times 30 \times (30 - 25.98)N = 947.19\text{N}$$

Hence, the answer is 947.19 (C).

Q15. Capillary waves are generated in the sea. The speed of propagation (C) of these waves is known to be a function of density (ρ), wave length (λ), and surface tension (σ). Assume, ρ and λ to be constant. If the surface tension is doubled, in the functional form of the relevant non-dimensional group, the percentage increase in propagation speed (C) is _______.

Solution:

Given:

Speed of propagation (C) of the wave is a function of density (ρ) wavelength (λ) and surface tension (σ)

$$i.e.\ C = f(\rho,\lambda,\sigma)$$

$$f(\rho,\lambda,\sigma,C) = 0$$

Here, total number of variable involved, m=4

And number of fundamental variable, n=3

∴ From Buckingham's pi theorem, the number of dimensionless terms=m-n=4-3=1

Now, dimensionless parameter can be written as,

$$\pi = \lambda^a \rho^b \sigma^c C \qquad \text{(i)}$$

where a, b and c are empirical constant sand λ,ρ and σ are repeating variables.

$$\text{Now, } \pi = L^a (ML^{-3})^b (MT^{-2})^c LT^{-1}$$

$$M^0 L^0 T^0 = M^{b+c} L^{a-3b+1} T^{-2c-1}$$

$$b + c = 0 \qquad \text{(ii)}$$

$$a - 3b + 1 = 0 \qquad \text{(iii)}$$

$$-2c - 1 = 0 \qquad \text{(iv)}$$

Now, from eq. (iv), we get, $c = -1/2$

Using values of c in eq. (ii), we get $b = 1/2$

And using values of b in eq. (iii), we get, $a = 1/2$

∴ Eq. (i) can be expressed as $\pi = \left(C\sqrt{\frac{\rho\lambda}{\sigma}}\right)$

$$C = \pi\left(\sqrt{\frac{\sigma}{\rho\lambda}}\right)$$

Now, assuming λ and ρ to be constant, we have

$$C = k\sqrt{\sigma}$$

Now, when σ is doubles, $C' = k\sqrt{2\sigma} = \sqrt{2}k\sqrt{\sigma}$

∴ Increase in speed of wave $= \sqrt{2}k\sqrt{\sigma} - k\sqrt{\sigma} = \left(\sqrt{2}-1\right)k\sqrt{\sigma}$

$\therefore$ The percentage increase in wave speed $=\dfrac{\left(\sqrt{2}-1\right)k\sqrt{\sigma}}{k\sqrt{\sigma}}\times 100\% = 41.42\%$

Hence, the answer is 41.12.

Q16. Consider a fully developed, two-dimensional and steady flow of a viscous fluid between two fixed parallel plates separated by a distance of 30 mm. The dynamic viscosity of the fluid is 0.01 kg/m-s and the pressure drop per unit length is 300 Pa/m. The fluid velocity at a distance of 10 mm from the bottom plate, in m/s, is ________.

Solution:

Given:

Distance between plates, h= 30 mm = 0.03m

Dynamic viscosity of fluid, μ = 0.01 kg/m-s

Pressure drop per unit length, $-\dfrac{dp}{dx} = 300\,\text{Pa/m}$

Now, velocity profile for flow between two fixed plates is given

$$]u = \frac{h^2}{2\mu}\left(-\frac{dp}{dx}\right)\left(\frac{y}{h}\right)\left(1-\frac{y^2}{h^2}\right) = \frac{(0.03)^2}{2\times 0.01}\times 300\times\left(\frac{0.01}{0.03}\right)\left[1-\frac{(0.01)^2}{(0.03)^2}\right]\text{m/s} = 3\,\text{m/s}$$

Hence, the answer is 3 m/s.

Q17. A 2.6 gram smooth table-tennis (ping-pong) ball has a diameter of 38 mm. Density (ρ) of air is 1.2 kg/m³. Neglect the effect of gravity. Take coefficient of drag as 0.5. If the ball is struck with an initial velocity of 30 m/s, the initial deceleration, in m/s², is ________.

Solution:

Given:

Diameter of the ball, D=38 mm=0.038 m

Mass of the ball, m = 2.6 g = 2.6 $\times 10^{-3}$ kg

Density of air, ρ = 1.2 kg/m^3

Coefficient of drag, C_D = 0.5

Initial velocity of ball, V = 30 m/s

Now, drag force on ball is given by,

$$F_D = \frac{1}{2}C_D\rho A V^2$$

$$= \frac{1}{2}\times 0.5\times 1.2\times\frac{\pi D^2}{4}\times 30\times 30$$

$$=\frac{1}{2}\times 0.5\times 1.2\times \frac{3.14\times (0.038)^2}{4}\times 30\times 30\text{ N}=0.3061\text{ N}$$

Now, we have F_D = m.a, where a is initial deceleration

$$\therefore 0.3061 = 2.6 \times 10^{-3} \times a$$

$$a = 117.7 \text{ m/s}^2$$

Hence, answer is 117.7.

Q18. On a flat plate, transition from laminar to turbulent boundary layer occurred at a critical Reynolds number (Re_{cr}). The empirical relations for the laminar and turbulent boundary layer thickness are given by $\frac{\delta_{lam}}{x}=5.48R_{ex}^{-0.5}$ and $\frac{\delta_{turb}}{x}=0.37R_{ex}^{-0.2}$ respectively. The ratio of laminar to turbulent boundary layer thickness, at the location of transition, is 0.3. The value of Re_{cr} is _______.

Solution:

Given:

$$\frac{\delta_{lam}}{x}=5.48R_{ex}^{-0.5}$$

$$\frac{\delta_{turb}}{x}=0.37R_{ex}^{-0.2}$$

$$\text{Now,}\frac{\frac{\delta_{lam}}{x}}{\frac{\delta_{turb}}{x}}=\frac{5.48R_{ex}^{-0.5}}{0.37R_{ex}^{-0.2}}$$

$$\frac{\delta_{lam}}{\delta_{turb}}=14.81R_{ex}^{-0.3}$$

$$14.81R_{ex}^{-0.3}=0.3$$

$$R_{ex}^{-0.3}=\frac{0.3}{14.81}=0.02025$$

$$R_{ex}=R_{cr}=(0.02025)^{-\frac{1}{0.3}}=441346.$$

Hence, the answer is 441346.15.

Q19. In a capillary tube of radius R = 0.25 mm, a fully developed laminar velocity profile is defined as, $u\times\frac{R^2}{4\mu}\left(-\frac{dp}{dx}\right)\left(1-\frac{r^2}{R^2}\right)$. In this expression, $-\frac{dp}{dx}=1\text{MPa/m}$, μ is the dynamic viscosity of the fluid, and r is the radial position from the centerline of the tube. If the flow rate through the tube is 1000 mm³/s, the viscosity of the fluid, in Pa–s, is ____________.

Solution:

Given:

$$R = 0.25\text{mm} = 0.25 \times 10^{-3}\text{m}$$
$$-\frac{dp}{dx} = 1\,MPa/m = 10^6\,\text{Pa/m}$$

Flow rate, Q = 1000 mm³/s = 1000×10⁻⁹ m³/s = 10⁻⁶ m³/s

Now, from continuity equation we have

$Q = A \times V$

$Q = \pi R^2.V$

$V = Q/\pi R^2$

$V = 5.092$ m/s

∴The average velocity of flow $V = 5.092$ m/s

Now, average velocity is given by,

$$V = \frac{R^2}{8\mu}\left(-\frac{dp}{dx}\right)$$
$$5.092 = \frac{(0.25 \times 10^{-3})^2}{8\mu} \times 10^6$$
$$\mu = \frac{(0.25 \times 10^{-3})^2}{8 \times 5.092} \times 10^6\,\text{Pa.s} = 0.0015\text{Pa.s}$$

∴The dynamic viscosity of the fluid is 0.0015 Pa.s

Hence, the answer is 0.0015

Q20. The skin friction coefficient for a turbulent pipe flow is defined as, $C_f = \frac{\tau_w}{1/2\rho V^2}$ where τ_w is the wall shear stress and V is the average flow velocity. The value of C_f is empirically given by the relation: $C_f = 0.065\left(2/Re\right)^{0.25}$, where *Re* is the Reynolds number. If the average flow velocity is 10 m/s, diameter of the pipe is 250 mm, kinematic viscosity of the fluid is 0.25×10⁻⁶ m²/s, and density of the fluid is 700 kg/m³, the skin friction drag induced by the flow over 1 m length of the pipe, in N, is __________.

Solution:

Given:

Average velocity of flow, V= 10 m/s

Diameter of pipe, d = 250 mm=0.25 m

Kinematic viscosity of fluid, $\nu = 0.25 \times 10^{-6}$ m²/s

Density of fluid, $\rho = 700$ kg/m^3

∴ The dynamic viscosity of the fluid $\mu = \rho\nu = 0.25\times10^{-6} \times 700$ Pa.s = 0.000175 Pa.s

The Reynolds number is given by

$$Re = \frac{\rho Vd}{\mu}$$
$$= \frac{700 \times 10 \times 0.25}{0.000175}$$
$$= 10000000$$

Now we have $C_f = 0.065\left(\frac{2}{Re}\right)^{0.25}$

$$= 0.065 \times \left(\frac{2}{10000000}\right)^{0.25}$$
$$= 0.001374$$

Also, we have $C_f = \frac{\tau_w}{\frac{1}{2}\rho V^2}$

$$\tau_w = \frac{1}{2} C_f \rho V^2$$

$$\frac{F}{A} = \frac{1}{2} C_f \rho V^2$$

$$F = \frac{1}{2} C_f \rho A V^2$$

$$F = \frac{1}{2} C_f \rho \pi d L V^2$$

$$F = \frac{1}{2} \times 0.001374 \times 700 \times 3.14 \times 0.25 \times 1 \times 10 \times 10 = 37.78N$$

Hence, answer is 37.78

Q21. A (150 mm × 150 mm) square pillar is located in a river with water flowing at a velocity of 2 m/s, as shown in figure. The height of the pillar in water is 8 m. Take density of water as 1000 kg/m^3 and kinematic viscosity as 1×10^{-6} m/s^2. The coefficient of drag of the pillar is 2.0. The drag force exerted by water on the pillar in N is _______.

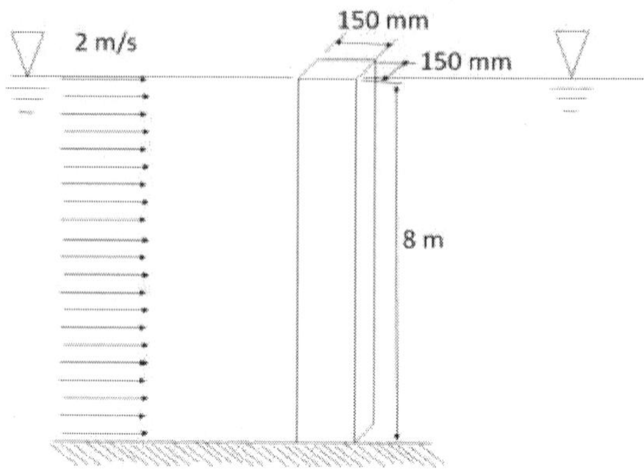

Solution:

Given:

Velocity of fluid flow, V= 2 m/s

Height of the pillar= 8 m

Density of water, ρ =1000 kg/m^3

Kinematic viscosity of fluid, $\nu = 1 \times 10^{-6}$ m^2/s

Coefficient of drag, $C_D = 2$

Now, the force exerted by water on the pillar is given by

$$F = \frac{1}{2} C_D \rho A V^2$$

$$= \frac{1}{2} \times 2 \times 1000 \times 8 \times 0.15 \times 2 \times 2\,\text{N} = 4800\,\text{N}$$

Hence, the answer is 4800.

Q22. An orifice plate is used to measure flow rate of air (density = 1.23 kg/m^3) in a duct of 250 mm diameter as shown in figure. The volume flow rate is 1 m^3/s. Flow at sections 1 and 3 is uniform and section 2 is located at vena contracta. The diameter ratio, D_t/D_1, is 0.66. The flow area at vena contracta, $A_2 = 0.65A_t$, where A_t is area of the orifice. The pressure difference between locations 2 and 3 in N/m^2 is __________.

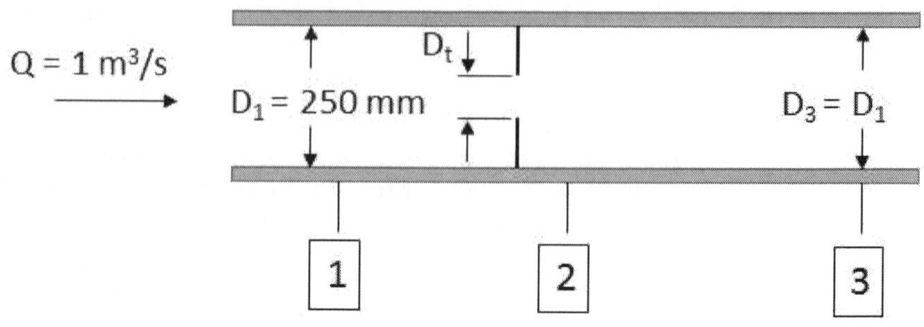

Solution:

Given:

Density of air, ρ =1.23 kg/m^3

Diameter of duct, D_1=D_3=250 mm=0.25 m

Volumetric flow rate, Q= 1 m^3/s

Diameter ratio, D_t/D_1 = 0.66.

Flow area at vena contracta, A_2=0.65 A_t

Hence the flow velocity at section $1, V_1 = V_3 = \frac{4Q}{\pi D_1^2} = \frac{4\times1}{\pi\times0.25}\text{m/s} = 20.37\ \text{m/s}$

Now, from the continuity equation we have, $A_1 V_1 = A_t V_t$

$$\frac{\pi D_1^2}{4}V_1 = \frac{\pi D_t^2}{4}V_t$$

$$D_1^2 V_1 = D_t^2 V_t$$

$$\frac{V_1}{V_t} = \left(\frac{D_t}{D_1}\right)^2 = (0.66)^2 = 0.4356$$

$$V_t = \frac{20.37}{0.4356}\text{m/s} = 46.763\text{m/s}$$

Again, from the continuity equation,

$A_2 V_2 = A_t V_t$

$0.65\ A_t V_2 = A_t \times 46.763$

$V_2 = 71.94$ m/s

Now, applying Bernoulli's equation between point 2 and 3

$$\frac{p_2}{\rho g} + \frac{V_2^2}{2g} + z_2 = \frac{p_3}{\rho g} + \frac{V_3^2}{2g} + z_3$$

$$p_3 - p_2 = \frac{\rho\left(V_2^2 - V_3^2\right)}{2}$$

$$p_3 - p_2 = \frac{1.23\times\{(71.94)^2 - (20.37)^2\}}{2}\text{Pa} = 2927.66\text{Pa}$$

Hence, the answer is 2927.66 Pa.

GATE 2017
FLUID MECHANICS

Q1. In the given flow field, the velocity vector in Cartesian coordinate system is given as:

$$\vec{V} = \left(x^2 + y^2 + z^2\right)\hat{i} + \left(xy + yz + y^2\right)\hat{j} + \left(xz - z^2\right)\hat{k}$$

What is the volume dilation rate of the fluid at a point where $x = 1$, $y = 2$ and $z = 3$?

(A) 6 (B) 5 (C) 10 (D) 0

Solution:

Volume Dilation rate is the rate of increase of volume of a fluid element per unit volume. Therefore, the volume dilation rate is given by

$$\frac{1}{\mathcal{V}}\frac{\partial \mathcal{V}}{\partial t} = \frac{\partial u}{\partial x} + \frac{\partial v}{\partial y} + \frac{\partial w}{\partial z}$$

Now for the given velocity field:

$$\frac{\partial u}{\partial x} = 2x$$

$$\frac{\partial v}{\partial y} = x + z + 2y$$

$$\frac{\partial w}{\partial z} = x - 2z$$

Therefore,

$$\frac{\partial u}{\partial x} + \frac{\partial v}{\partial y} + \frac{\partial w}{\partial z} = 2x + x + z + 2y + x - 2z$$

$$\frac{\partial u}{\partial x} + \frac{\partial y}{\partial x} + \frac{\partial w}{\partial z} = 4x + 2y - z$$

At $x = 1, y = 2, z = 3$

$$\frac{\partial u}{\partial x} + \frac{\partial v}{\partial y} + \frac{\partial w}{\partial z} = 5$$

Hence, the answer is (B).

Q2. A steady, incompressible, two-dimensional velocity field in Cartesian coordinate system is represented by the following expression:

$$\vec{V} = (0.7 + 0.4x)\hat{\imath} + (1.2 - 0.4y)\hat{\jmath}$$

The coordinates of the point (x, y) in the flow field having "zero" velocity is

(A) (1.75, -3) (B) (-1.75, 3) (C) (1.75, 3) (D) (-1.75, -3)

Solution:

Given: $\vec{V} = (0.7 + 0.4x)\hat{\imath} + (1.2 - 0.4y)\hat{\jmath}$

Now, zero velocity means, $\vec{V} = 0\hat{\imath} + 0\hat{\jmath}$

i.e., $0.7 + 0.4x = 0$ and $1.2 - 0.4y = 0$

$0.4x = -0.7$ and $0.4y = 1.2$

$x = -1.75$ and $y = 3$

Hence, the answer is (B).

Q3. During an experiment, the position of a fluid particle is monitored by an instrument over a time period of 10s. The trace of the particle given by the following figure represents a

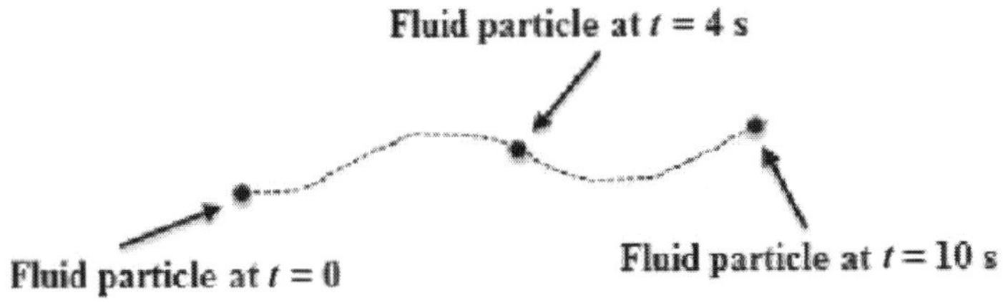

(A) Streamline (B) Streakline

(C) Pathline (D) Timeline

Solution:

Streamline: stream line is the imaginary line in the flow field where tangent drawn at any point indicates the direction of the velocity at that point.

Streakline: Streakline is the instantaneous locus of all the fluid particles that have previously passed through a fixed point.

Pathline: Pathline is the path traced by a single fluid particle over a period.

Timeline: Timeline is the line obtained by marking a number of adjacent fluid particles at a given instant.

Hence, the answer is (C).

Q4. In a Cartesian two-dimensional coordinate system, u and v represent the velocities in x and y direction, respectively. For a certain flow, the velocity field is represented by the following expression:

$$\vec{V} = (ax + by)\hat{i} + (cx + dy)\hat{j}$$

where the coefficients a, b, c and d are constants. For an incompressible flow, which one of the following relation is TRUE?

(A) a + d=0 (B) a + c=0 (C) b + d=0 (D) b + c=0

Solution:

We know that for an incompressible flow

$\nabla.\vec{V} = 0$ (Continuity equation in vector form)

$$\nabla.\{(ax + by)\hat{i} + (cx + dy)\hat{j}\} = 0$$

$$\frac{\partial(ax + by)}{\partial x} + \frac{\partial(cx + dy)}{\partial x} = 0$$

$a + d = 0$

Hence, the answer is (A).

Q5. Which of the following figures represents potential flow past a circular cylinder with clockwise rotation of the cylinder?

(A)

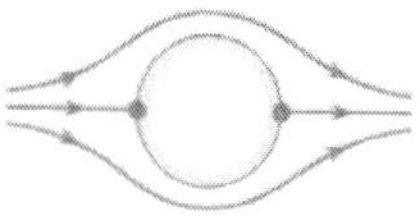

(B)

(C)

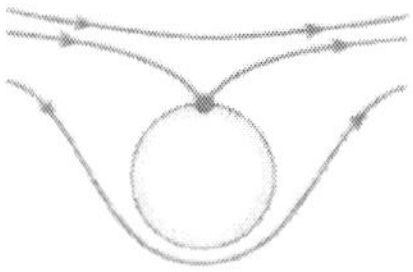

(D)

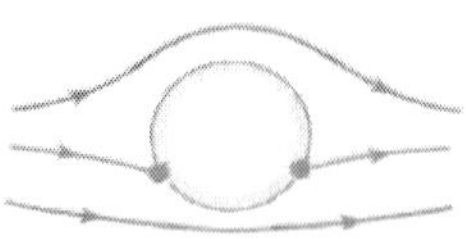

Solution:

(A) represents the potential flow over a stationary cylinder.

(B) represents the potential flow over a circular cylinder with anticlockwise rotation when sinα<1.

(C) representsthe potential flow over a circular cylinder with anticlockwise rotation when sinα=1.

(D) represents the potential flow over a circular cylinder with clockwise rotation when sinα<1.

Hence, the answer is (D).

Q6. The steam function (Ψ) of velocity field at any location (x, y) is given as $\Psi = xy^2 - 2x^2y^2$. What is the rate of rotation of a fluid element located at (2, 2)?

(A) 8 (B) 10 (C) 12 (D) 14

Solution:

Given: $\Psi = xy^2 - 2x^2y^2$

We have, $\frac{\partial \psi}{\partial x} = y^2 - 4xy^2$ and $\frac{\partial \psi}{\partial y} - 2xy - 4x^2y$

Now, for a steam function Ψ

$$\frac{\partial \psi}{\partial x} = -v \quad \text{and} \quad \frac{\partial \psi}{\partial x} = u$$

$$y^2 - 4xy^2 = -v \quad \text{and} \quad 2xy - 4x^2y = u$$

$$v = 4xy^2 - y \quad \text{and} \quad u = 2xy - 4x^2y$$

$$\frac{\partial v}{\partial x} = 4y^2 \quad \text{and} \quad \frac{\partial u}{\partial y} = 2x - 4x^2$$

Now, rotational component is given by

$$w_z = \frac{1}{2}\left(\frac{\partial v}{\partial x} - \frac{\partial u}{\partial y}\right)$$

$$= \frac{1}{2}\left(4y^2 - 2x + 4x^2\right)$$

Hence the angular velocity at the point ($x = 2, y = 2$)

$$w_z = \frac{1}{2}\left(4*2^2 - 2*2 + 4*2^2\right) = \frac{1}{2}(16 - 4 + 16) = 14$$

Hence, the answer is (D).

Q7. The nature of velocity profile within the laminar viscous sublayer in a turbulent pipe flow is

(A) Linear (B) Parabolic (C) Logarithmic (D) Exponential

Solution:

In boundary layer region, there are three regimes viz. laminar, transition and turbulent. Within the turbulent boundary layer, another regime exists with negligible thickness called sublayer.

Velocity profile in different regimes.

Laminar: parabolic

Turbulent: exponential or logarithmic.

Sublayer: actually, velocity profile in the sublayer is parabolic but due to its negligible thickness, it is assumed that velocity profile is almost linear.

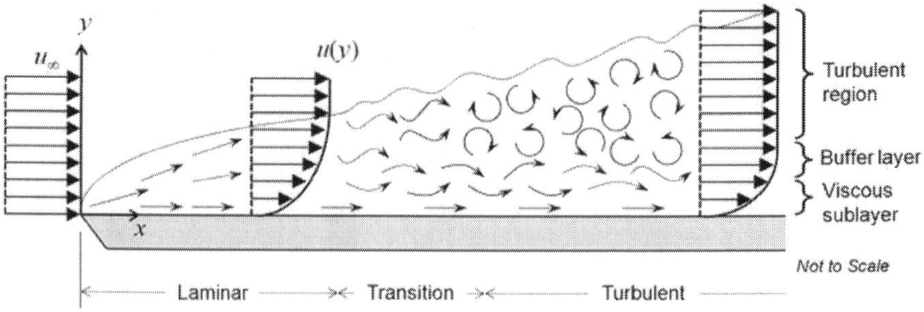

Hence, the answer is (A).

Q8. In a 5 m deep vertical cylindrical tank, water is filled up to a level of 3 m from the bottom of the tank and the remaining space is filled with oil of specific gravity 0.88. Assume density of water 1000kg/m^3 and acceleration due to gravity to be 10m/s^2. The gauge pressure (in kN/m^2, round off to the first decimal place) at a depth of 2.5m from the top of the tank will be ______.

Solution:

Total pressure is given by

$p = \rho_1 gh_1 + \rho_2 gh_2$

$= (0.88 \times 1000) \times 10 \times 2 + 1000 \times 10 \times 0.5$

$= 22600$ N/m^2

$= 22.6$ kN/m^2

Hence, the answer is 22.6 kN/m^2.

Q9. In a two dimensional potential flow, a point source at the origin ($x = 0$, $y = 0$) as shown in the figure. The strength of the point source is 2 cm^2/s. A uniform flow with the velocity 1 cm/s is approaching towards the point source at an angle of 30° from the horizontal axis. What is the distance (cm) of the stagnation point in the flow from the point source?

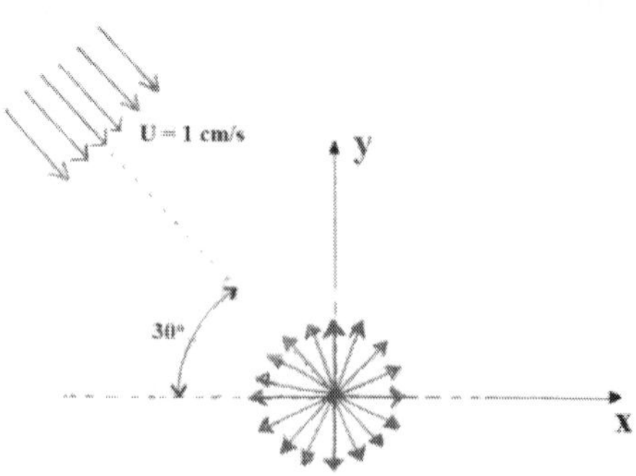

(A) $1/\pi$ (B) $2/\pi$ (C) $1/2\pi$ (D) $\sqrt{3}/2\pi$

Solution:

We know that the potential field,

$$\psi = Ur\sin\theta + \frac{q\theta}{2\pi}$$

$$\frac{\partial\psi}{\partial\theta} = Ur\cos\theta + \frac{q}{2\pi}$$

$$\frac{\partial\psi}{r\partial\theta} = U\cos\theta + \frac{q}{2r\pi}$$

$$v_r = v_x = \frac{\partial\psi}{r\partial\theta} = U\cos\theta + \frac{q}{2r\pi} \quad \text{(i)}$$

At, the stagnation point, $\theta = \frac{5\pi}{6}, v_r = 0, x_s = -r\cos\frac{\pi}{6}$

Putting the values in eq. (i), we get

$$U\cos\frac{5\pi}{6} - \frac{q}{2\pi x_s}\cos\frac{5\pi}{6} = 0$$

$$x_s = \frac{q}{2\pi U}$$

$$x_s = \frac{2}{2\pi\times 1}\text{cm} = \frac{1}{\pi}$$

Hence, the answer is (A).

Q10. Two infinite parallel plates are separated by a small gap (d = 20 mm) as shown in figure. The bottom plate is fixed and the gap between the plates is filled with oil having density of 890 kg/m^3 and kinematic viscosity of 0.00033 m^2/s. A shear force is induced by moving the upper plate with a velocity of 5m/s. Assume linear velocity profile between the plates and the oil to be a Newtonian fluid. The shear stress (N/m^2) at the upper plate is _______.

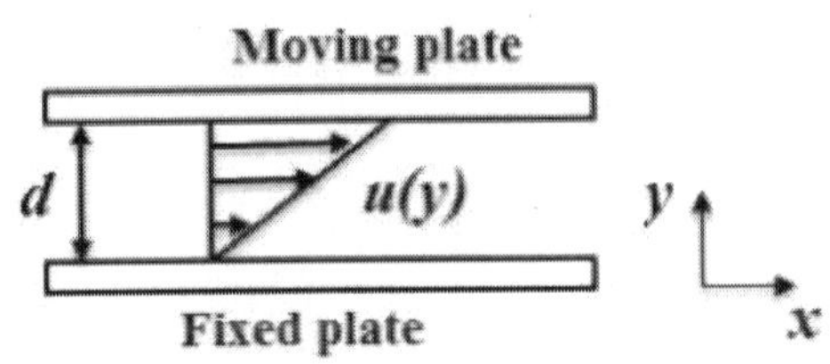

Solution:

Given:

$dy = 20 \times 10^{-3}$ m

$du = u - 0 = (5 - 0)$ m/s = 5 m/s

$\nu = 0.00033$ m²/s

$\rho = 890$ kg/m³

Therefore, $\mu = \rho \times \nu$

= 0.00033 × 890

= 0.2937 kg/ms

Now, shear stress, $\tau = \mu \dfrac{du}{dy} = 0.2937 * \dfrac{5}{20 * 10^{-3}} = 73.425\,\text{N/m}^2$

Hence, the answer is 73.425.

Q11. A spherical balloon of diameter 15 m is supposed to lift a load of 3000 N. The lifting of load is achieved by heating the air inside the balloon. Assume, air to be an ideal gas and atmospheric pressure either outside or inside the balloon. The value of acceleration due to gravity is 9.81 m/s² and the values of temperature and density of atmospheric air are 15°C and 1.2 kg/m³, respectively. In order to lift the specified load, the air inside the balloon should be heated to a temperature of ______.

Solution:

Given:

Diameter of the balloon, $d = 15$ m

Force to be lifted, $F = 3000$ N

Initial Temperature, T_1= 15°C i.e.288 K Density, ρ_1= 1.2 kg/m³

Let final temperature be T_2, final density be ρ_2 and V be the volume of balloon.

Now, when air inside the balloon is heated to T_2, the forces acting on the balloon are:

Weight given by, $W= \rho_2 \forall g$ buoyancy force given by, $F_b= \rho_1 \forall g$ and force to be lifted, $F = 3000$ N

Now, in order to lift the force F, the buoyancy force must be equal to the sum of weight and force to be lifted, F.

i.e., $F_b = W + F$

$\rho_1 Vg = \rho_2 Vg + 3000$

$1.2 \times \pi d^3/6 \times 9.81 = \rho_2 \times \pi d^3 \times 9.81 + 3000$

$1.2 \times 3.14 \times 15^3/6 \times 9.81 = \rho_2 \times 3.14 \times 15^3/6 \times 9.81 + 3000$

$\rho_2 = 1.0268$ kg/m³

Now, for ideal gas at constant pressure, $\rho_1 T_1 = \rho_2 T_2$

$T_2 = \rho_1 T_1/\rho_2$

$T_2 = 1.2 \times 288/1.0268$

T_2= 336.57 K= 63.57

Hence, the answer is 63.57.

Q12. The velocity field in Cartesian coordinate system for a two-dimensional system steady flow is given as:

$$\vec{V} = \left(\frac{v_0}{L}\right)\left(x\hat{i} - y\hat{j}\right)$$

where, v_0 and L are constants. Which of the following represents acceleration field (a) for the flow?

(A) a = 0

(B) $a = \left(\frac{v_0}{L}\right)(xi - yj)$

(C) $a = \left(\frac{v_0^2}{L^2}\right)(xi - yj)$

(D) $a = \left(\frac{v_0^2}{L^2}\right)(xi - yj)$

Solution:

Given:

$$\vec{V} = \left(\frac{v_0}{L}\right)\left(x\hat{i} - y\hat{j}\right)$$

$$= \frac{v_o}{L} x_{\hat{i}} + \left(-\frac{v_0}{L} y\right)\hat{j}$$

$$u = \frac{v_0}{L} x \qquad \text{and} \qquad v = -\frac{v_0}{L} y$$

$$a_x = u\frac{\partial u}{\partial x} + v\frac{\partial u}{\partial y} \quad \text{and} \quad a_y = v\frac{\partial v}{\partial x} + v\frac{dv}{dy}$$

$$= \left(\frac{v_0}{L} x\right)\frac{v_0}{L} + 0 \qquad = 0 + \left(-\frac{v_0}{L} y\right)\left(-\frac{v_0}{L}\right)$$

$$= \frac{v_0^2}{L^2} x \qquad = \frac{v_0^2}{L^2} y$$

Therefore, acceleration field

$$a = a_x \hat{i} + a_y \hat{j}$$

$$= \frac{v_0^2}{L^2} x\hat{i} + \frac{v_0^2}{L^2} y\hat{j}$$

$$= \left(\frac{v_0^2}{L^2}\right)\left(x\hat{i} + y\hat{j}\right)$$

Hence, the answer is (D).

Q13. A cylinder tank of 0.8 m diameter is completely filled with water and its top surface is open to atmosphere as shown in the figure. Water is being discharged to the atmosphere from a circular hole of 15 mm diameter located at the bottom of the tank. The value of acceleration due to gravity is 9.81 m/s². How much time (s) would be required for water level to drop from a height of 1m to 0.5 m?

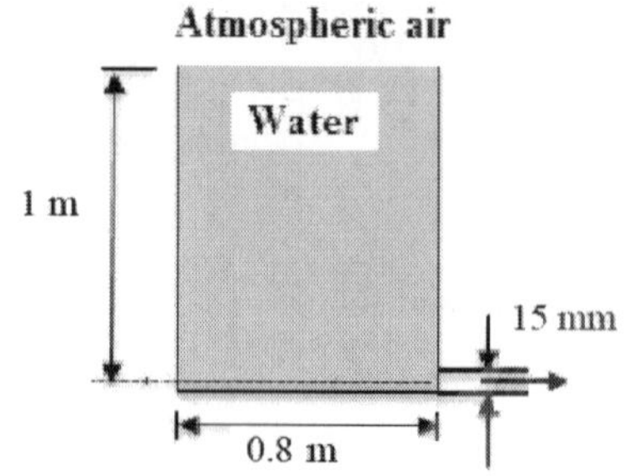

(A) 188 (B) 266 (C) 376 (D) 642

Solution:

Given:

Diameter of the tank, d_1= 0.8 m

Diameter of the pipe, d_2= 15 mm (0.015m)

Now, let at any time *t*, water level in the tank is *h* and in small time *dt*, water level in the tank decreases by *dh*.

Therefore, flow rate in tank is given by, $Q_1 = \frac{\pi d_1^2}{4}\times\left(\frac{-dh}{dt}\right)$

And flow rate in the pipe is given by, $Q_1 = \frac{\pi d_1^2}{4}\times\sqrt{2gh}$

Now, from continuity equation, $Q_1 = Q_2$

$$\frac{\pi d_1^2}{4}\times\left(\frac{-dh}{dt}\right)=\pi\frac{\pi d_2^2}{4}\times\sqrt{2gh}$$

$$0.8^2\times\frac{-dh}{dt}=0.015^2\times\sqrt{2*9.81}*\sqrt{h}$$

$$642.166\times\left(\frac{-dh}{dt}\right)=\sqrt{h}$$

$$dt=642.166\times\left(\frac{-dh}{\sqrt{h}}\right)$$

$$t=-642.166\int_1^{0.5}h^{-0.5}dh$$

$$t=642.166\times0.5857864\,\text{s}$$

$$t=376.17\,\text{s}$$

Hence, the answer is (C).

Q14. Consider steady laminar flow of an incompressible Newtonian fluid between two infinite parallel plates separated by a distance of 1m as shown in fig. The bottom plate is stationary but the top one is moving in positive X direction with a velocity of 3m/s. The fluid pressure gradient in the flow direction $\frac{\partial p}{\partial x} = -18\,\text{N/m}$. If the viscosity of the fluid is 1 kg/ms then the distance of the point of maximum velocity (in meters, rounded off to the second decimal place) from the bottom plate would be

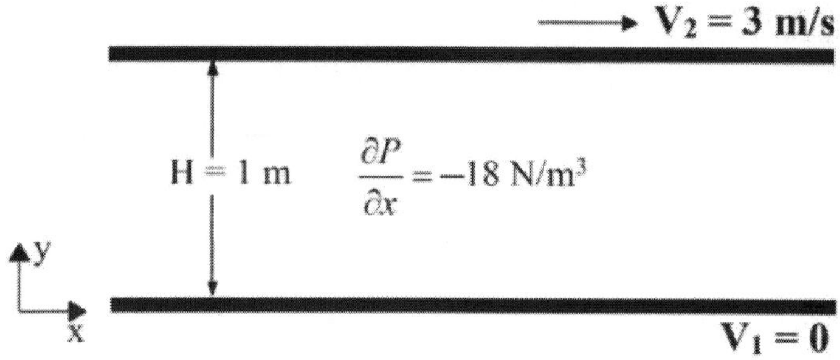

Solution:

For Couette flow, a dimensionless form of pressure gradient is defined as:

$$\alpha = \frac{h^2}{2\mu U}\left(-\frac{dp}{dx}\right)$$

where, h = distance between the plate

μ = dynamic viscosity of the fluid

U = velocity of the top plate

$\frac{dp}{dx}$ = pressure gradient

And, velocity profile for Couette flow is given by

$$\frac{u}{U} = \frac{y}{h} + \alpha\frac{y}{h}\left(1 - \frac{y}{h}\right)$$

Now, for maxima or minima, we have

$$\frac{1}{U}\frac{du}{dy} = \frac{1}{h} + \alpha\left(\frac{1}{h} - 2\frac{y}{h^2}\right) = 0$$

$$\alpha\left(\frac{1}{h} - 2\frac{y}{h^2}\right) = -\frac{1}{h}$$

$$y = \left(\frac{\alpha + 2}{2ga}\right)h$$

Now, $U = 3$ m/s

$$-\frac{dp}{dt} = 18\,\text{N/m}^2$$

$h = 1$ m and $\mu = 1$ kg/ms

Therefore, $\alpha = \frac{h^2}{2\mu U}\left(\frac{-dp}{dx}\right)$

$$\alpha = \frac{1^2}{2\times1\times3}\times18 = 3$$

Therefore the maximum velocity will occur at $y = \left(\frac{\alpha+1}{2\alpha}\right)h$

$$y = \left(\frac{3+1}{2\times3}\right)\times1 = 0.67$$

Hence, the answer is 0.67.

Q15. An inviscid incompressible fluid density 1000 kg/m³ is flowing in a horizontal pipe of tapered cross-section with a flow rate of 4000 cm³/s. The area of cross-section at two different locations '*A*' and '*B*' are 10 cm² and 20 cm², respectively. The velocity of the fluid at the location '*A*' is 4 m/s and pressure is 5 N/m². The pressure at location '*B*' would be

Solution:

Given:

Density, ρ =1000kg/m³

Flow rate, Q = 4000 cm³/s = 0.004 m³/s

A_1=0.001m²

V_1=4 m/s

p_1=5 N/m²

A_2=0.002 m²

From the equation of continuity, we have, $A_1V_1 = A_2V_2$

$$V_2 = \frac{A_1\times V_1}{A_2} = \frac{0.001\times4}{0.002} = 2\,\text{m/s}$$

From Bernoulli's equation

Since, the pipe is horizontal so, $z_1 = z_2$

$$\frac{p_1}{\rho g} + \frac{v_1^2}{2g} + z_1 = \frac{p}{\rho g} + \frac{V_2^2}{2g} + z_2$$

$$\frac{5}{1000 \times g} + \frac{4^2}{2g} = \frac{p_2}{1000 \times g} + \frac{2^2}{2g}$$

Solving the equation, we get, $p_2 = 6005\ N/m^2$

Hence, the answer is 6005.

Q16. A viscous, incompressible and Newtonian fluid flowing through the main branch of a circular pipe bifurcates into two daughter branches whose radii are 4 cm and 2 cm, respectively. The flow in both the daughter branches are laminar and fully developed. If the pressure gradients in both the daughter branches are same, the fraction of total volumetric flow rate (rounded off to the second decimal place) coming out from the branch with 4 cm diameter is _______.

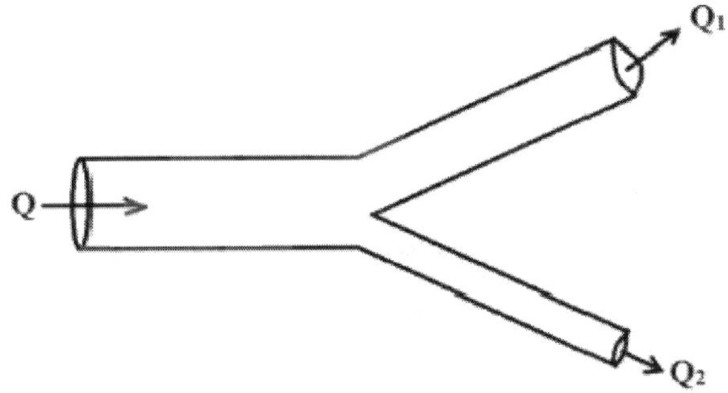

Solution:

Given, radii are $r_1 = 4$ cm and $r_2 = 2$ cm,

And we have to find Q_1/Q=?

The average velocity in the branch with radius r_1 is

$$\bar{v}_1 = -\frac{r_1^2}{8\mu}\frac{dp}{dz}$$

Hence flow rate through the pipe with radius r_1 is

$$Q_1 = \pi r_1^2 \bar{v}_1 = -\frac{\pi r_1^4}{8\mu}\frac{dp}{dz}$$

Similarly flow rate through the pipe with radius r_2 is

$$Q_2 = \pi r_2^2 \bar{v}_1 = -\frac{\pi r_2^4}{8\mu}\frac{dp}{dz}$$

Hence $\frac{Q_2}{Q} = \frac{Q_2}{Q_1 + Q_2} = \frac{r_2^4}{r_1^4 + r_2^4} = \frac{2^4}{4^4 + 2^4} = 0.06$

Hence, the answer is 0.06.

Q17. The volumetric flow rate (Q) of a triangular notch is a function of the upstream liquid surface elevation (H) measured from the bottom of the notch, acceleration due to gravity (g), notch angle (φ) and the approach velocity (V). Which one of the following is the correct expression for Q?

(A) $Q = H^{\frac{1}{2}} * f\left(\frac{v}{\sqrt{H}}, \sqrt{g\phi}\right)$ (B) $Q = H * f\left(\frac{v}{\sqrt{H}}, \sqrt{g\phi}\right)$

(C) $Q = H^{\frac{3}{2}} * f\left(\frac{v}{\sqrt{H}}, \sqrt{g\phi}\right)$ (D) $Q = H^{\frac{5}{2}} * f\left(\frac{v}{\sqrt{H}}, \sqrt{g\phi}\right)$

Solution:

$$Q = (H, g, \emptyset, v)$$

$$f_1 (Q, H, g, \emptyset, v) = 0 \qquad \text{(i)}$$

Using Buckingham's π-Theorem

Hence the total number of variables, n=5 Dimensions of each variable

$Q = L^3T^{-1}$ $H = L^1$ $g = L^1T^{-2}$ φ=Dimensionless $v = L^1T^{-1}$

Hence the total number of fundamental variable, m=2

So the number of π terms = n-m= 5-2=3

Each π term contains m +1 variables,

The repeating variables are H, g, so we have

$\pi_1 = H^{a1}g^{b1}Q$

$\pi_2 = H^{a2}g^{b2}v$

$\pi_3 = H^{a3}g^{b3}\emptyset$

First π term, $\pi_1 = H^{a1}g^{b1}Q$

Substituting dimensions on both side of π_1,

$M^0L^0T^0 = [L^1]^{a1}[L^1T^{-2}]^{b1}[L^3T^{-1}]$

Equating the powers of M, L, T on both sides,

We get, $a1 = -\frac{5}{2}, b1 = -\frac{1}{2}$

$\pi_1 = H^{-\frac{5}{2}}g^{\frac{1}{2}}Q$

$\pi_1 = \frac{Q}{H^{\frac{5}{2}} * \sqrt{g}}$

Second π term, $\pi_2 = H^{a2}g^{b2}v$

Substituting dimensions on both side of π_2,

$M^0L^0T^0 = [L^1]^{a2}[L^1T^{-2}]^{b2}[L^1T^{-1}]$

Equating the powers of M, L, T on both sides,

We get, $a2 = -\frac{-1}{2}, b1 = -\frac{1}{2}$

$$\pi_2 = H^{\frac{1}{2}} g^{\frac{1}{2}} v$$

$$\pi_2 = \frac{v}{\sqrt{H}\sqrt{g}}$$

Third π term, $\pi_3 = H^{a3} g^{b3} \emptyset$

Substituting dimensions on both side of π_3,

$M^0 L^0 T^0 = [L^1]^{a3} [L^1 T^{-2}]^{b3}$

Equating the powers of M, L, T on both sides,

We get, a3=0 and b3=0

$\pi_3 = H^0 g^0 \emptyset$

$\pi_3 = \emptyset$

Equation (i) can be written as $f_1 (\pi_1, \pi_2, \pi_3) = 0$

$$f_1\left(\frac{Q}{H^{\frac{5}{2}} * \sqrt{g}}, \frac{v}{\sqrt{H}\sqrt{g}}, \emptyset\right) = 0$$

$$\frac{Q}{H^{\frac{5}{2}} * \sqrt{5}} = f\left(\frac{v}{\sqrt{H}\sqrt{g}}, \emptyset\right)$$

$$Q = H^{\frac{5}{2}} * \sqrt{g} f\left(\frac{v}{\sqrt{H}\sqrt{g}}, \emptyset\right)$$

$$Q = H^{\frac{2}{5}} * f\left(\frac{v}{\sqrt{H}}, \sqrt{g}\emptyset\right)$$

Hence, the answer is (D).

Q18. Model tests are to be carried out to study the flow through a large prototype valve of 0.6 m diameter at a flow rate of 10 m³/s. The same working fluid is used for both the model and the prototype. A complete geometric similarity is maintained between the model and the prototype. If the valve diameter of the model is 80 mm, its required flow rate (in m³/s, rounded off to the first decimal place) would be

Solution:

Given:

Diameter of the prototype = 0.6 m

Diameter of the model = 0.08m

Flow rate of the prototype = 10 m^3/s

For the flow through pipe viscous force is the dominating force over other forces, therefore, Reynolds number for model and prototype will be equal.

i.e. $Re_p = Re_m$

$$\frac{pv_p d_p}{\mu} = \frac{pv_m d_m}{\mu}$$

$$v_m = \frac{v_p d_p}{d_m} \tag{i}$$

Now, $Q_p = 10$

$A_p v_p = 10$

$$v_p = \frac{10}{\frac{\pi}{4} * 0.6^2} m/s$$

$v_p = 35.37$ m/s

From eq. (i),

$$v_m = \frac{v_p d_p}{d_m} = \frac{35.37 * 0.6}{0.08} m/s = 265.275 m/s$$

Therefore, the flow rate through the model is,

$$Q_m = A_m v_m$$

$$= \frac{\pi}{4} * 0.08^2 * 265.275 m^3/s$$

$$= 1.3 m^3/s$$

Hence, the answer is 1.3 m^3/s.

Q19. Water is flowing at a rate of 0.5 m^3/s in a horizontal pipeline of inside diameter 0.5 m. The density and kinematic viscosity of water is 1000 kg/m^3 and 10^{-6} m^2/s, respectively. Assume Darcy-Weisbach friction factor value to be 0.0093 and acceleration due to gravity as 9.81 m/s^2. To maintain constant flow rate, the required power per unit length of the pipeline (in W/m, rounded off to the first decimal place) would be

Solution:

Given:

Volumetric flow rate, $Q = 0.5 m^3/s$

Diameter, $d = 0.5$m

Density, $\rho = 1000 kg/m^3$

Kinematic Viscosity, $\nu = 10^{-6} m^2/s$

Darcy-Weisbach friction factor, $f = 0.0093$

$g = 9.18 m/s^2$

Loss of head due to friction, $h_f = \dfrac{fLV^2}{2gd}$

$$v = \frac{Q}{A}$$

$$v = \frac{0.5}{\frac{\pi}{4}0.5^2} = \frac{8}{\pi}$$

$$\therefore h_f = \frac{0.0093 \times L \times \left(\frac{8}{\pi}\right)^2}{2 \times 9.81 \times 0.5}$$

$$= 0.00615L \text{ m}$$

Hence the power Required, $P = \rho \times g \times Q \times h_f$

P = 1000 × 0.5 × 9.81 × 0.00615L

$$\frac{P}{L} = 1000 \times 0.5 \times 9.81 \times 0.00615 \text{W/m}$$

$$= 30.2 \text{W/m}$$

Hence, the answer is 30.2 W/m.

Q20. Air flows over a smooth plate at a velocity of 4.39 m/s. The density of Air is 1.031 kg/m³ and the kinematic viscosity is 1.34 × 10⁻⁵ m²/s. The plate length is 12.2 m in the direction of flow. The boundary layer thickness (δ) is given as, where x is the distance from the leading edge $\frac{0.37x}{(\mathrm{Re}_x)^{\frac{1}{5}}}$, and Re_x is the Reynolds number. The boundary layer thickness (in meters, round off to the second decimal place) at 12.2 m distance from the leading edge will be _____.

Solution:

Here, $\mathrm{Re}_X = \dfrac{UL}{\vartheta}$

$$= \frac{4.39 \times 12.2}{1.34 \times 10^{-5}}$$

$$= 3.996 \times 10^6$$

Therefore, $\delta = \dfrac{0.37x}{(\mathrm{Re}_x)^{\frac{1}{5}}}$

$$= \frac{0.37 \times 12.2}{\left(3.996 \times 10^6\right)^{\frac{1}{5}}} 0.22m$$

Hence, the answer is 0.22.

Q21. A venturimeter of diameter 0.2 m at the entrance and 0.1 m at the throat is inclined upwards. The vertical elevation difference between the entrance and the throat is 0.5 m. The density of water is 1000 kg/m³. And the coefficient of velocity is 0.97. The differential U-tube manometer connected to the entrance and the throat shows a pressure difference of 30 kN/m². Assume acceleration due to gravity as 9.81m/s². The velocity of the water (in m/s, rounded off to first decimal place) at the throat would be

Solution:

Given:

At the entrance | At the throat

D_1=0.2 m | D_2=0.1 m

z_2-z_1=0.5 m

p_1-p_2=30 X 10³ N/m²

Coefficient of velocity = 0.97

ρ = 1000 kg/m³

Applying Bernoulli's Theorem,

$$\frac{p_1}{\rho g}+\frac{V_1^2}{2g}+z_1=\frac{p_2}{\rho g}+\frac{V_2^2}{2g}+z_2$$

$$\frac{30000}{1000g}-0.5=\frac{V_2^2}{2g}-\frac{V_1^2}{2g} \qquad (i)$$

From the equation of continuity, A_1V_1=A_2V_2

$$V_2=\frac{0.2^2}{0.1^2}V_1$$

$$V_2=4V_1$$

Putting the value of in eq. (i), we get

$$\frac{30000}{1000g}-0.5=\frac{15V_1^2}{2g}$$

Solving the equation, we get,

V_1 = 1.829 m/s

V_2 = 7.316 m/s

Required velocity, $(V_2)_{act}$ = 0.97 × 7.316 m/s = 7.09 m/s

Hence, the answer is 7.09 m/s.

Q22. A spherical bubble of radius r is rising upward with a constant velocity U, in quiescent water of dynamic viscosity μ. The density of air and water is denoted as ρ_a and ρ_w, respectively and g is the acceleration due to gravity. The bubble motion is such that, the Reynolds number Re<< 1. The density of air can be neglected in comparison to water ($\rho_a \ll \rho_w$). Which of the following relation is TRUE for density of water?

(A) $\rho_w = \frac{2}{9}\frac{\mu U}{r^2 g}$ (B) $\rho_w = \frac{2}{9}\frac{\mu U}{r^2 g}$

(C) $\rho_w = \frac{9}{4}\frac{\mu U}{r^2 g}$ (D) $\rho_w = \frac{4}{9}\frac{\mu U}{r^2 g}$

Solution:

Terminal velocity of a spherical particle with radius r and density ρ_p falling through a viscous medium having density ρ_m, viscosity μ provided Re < 0.2 is given by

Here, $\rho_m = \rho_a$

$$V_t = \frac{2}{9}\left(\frac{\rho_p - \rho_m}{\mu}\right)r^2 g$$

$V_t = U$

$\rho_p = \rho_w$

$$U = \frac{2}{9}\left(\frac{\rho_w - \rho_a}{\mu}\right)r^2 g$$

Now, since $\rho_a \ll \rho_w$ hence ρ_a can be neglected.

$$\rho_w = \frac{9}{2}\frac{\mu U}{r^2 g}$$

Hence, the answer is (B).

GATE 2016
FLUID MECHANICS

Q. 1 – Q. 9 carry one mark each.

Q1. In the parallel-plate configuration shown, steady-flow of an incompressible Newtonian fluid is established by moving the top plate with a constant speed, U_0 = 1 m/s. If the force required on the top plate to support this motion is 0.5 N per unit area (in m²) of the plate then the viscosity of the fluid between the plates is __________ N-s/m²

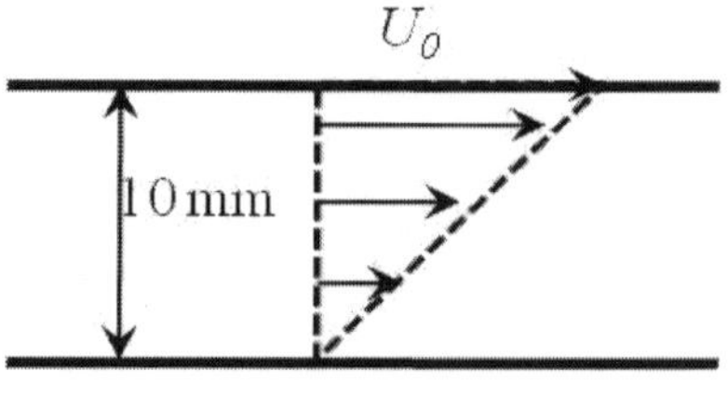

Solution:

We have, $\tau = \dfrac{F}{A}$ or, $\tau = \dfrac{0.5}{1} N/m^2 = 0.5\, N/m^2$

Now, $\tau = \mu \times \dfrac{du}{dy}$

or, $\mu \times \dfrac{du}{dy} = 0.5$

$\mu \times \dfrac{1}{0.01} = 0.5$ (Assuming a linear velocity profile)

$\mu = 0.5 \times 10^{-2}\, N/m^2 = 0.005\, N.s/m^2 q$

Hence, the answer is 0.005 N.s/m²

Q2. For a newly designed vehicle by some students, volume of fuel consumed per unit distance travelled (q_f in m³/m) depends upon the viscosity (μ) and density (ρ) of the fuel and, speed (U) and size (L) of the vehicle as

$$q_f = C\frac{\rho U^2 L}{\mu^3}$$

where C is a constant. The dimensions of the constant C are

(A) $M^0 L^0 T^0$ (B) $M^2 L^{-1} T^{-1}$ (C) $M^2 L^{-1} T^{-1}$ (D) $M^2 L^1 T^1$

Solution:

We have, $q_f = C\frac{\rho U^2 L}{\mu^3}$

$$C = \frac{qf.\mu^3}{\rho U^2 L}$$

$$= \frac{L^2.\left(ML^{-1}T^{-1}\right)^3}{ML^{-3}.\left(LT^{-1}\right)^2.L}$$

$$= M^2 L^{-1} T^{-1}$$

Hence, the answer is (B).

Q3. A semi-circular gate of radius 1m is placed at the bottom of a water reservoir as shown in figure below. The hydrostatic force per unit width of the cylindrical gate in *y*-direction is _______kN. The gravitational acceleration, g = 9.8 m/s² and the density of water = 1000 kg/m³.

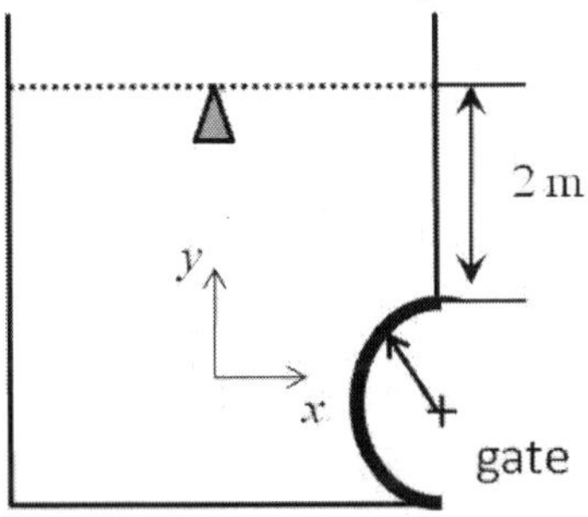

Solution:

Force in the *y*-direction = weight of the fluid inside the Semi-circular portion.

$$= \rho vg$$

$$= \rho\frac{\pi r^2 l}{2}g$$

$$= \frac{1000 \times \pi \times 1 \times 9.81}{2} N$$

$$= 15.4\,kN$$

Hence, the answer is 15.4 kN.

Q4. Velocity vector in m/s for a 2-D flow is given in Cartesian coordinate (x, y) as $\vec{V} = \left(\frac{x^2}{4}\hat{\imath} - \frac{xy}{2}\hat{j}\right)$. Symbols bear usual meaning. At a point in the flow field, the x- and y-components of the acceleration vector are given as 1 m/s^2 and -0.5 m/s^2, respectively. The velocity magnitude at that point is ________ m/s.

Solution:

For the given velocity field:

$$u = \frac{x^2}{4} \text{ and } v = -\frac{xy}{2}$$

Given that $a_x = 1$ and $a_y = -0.5$

$$a_x = u\frac{\partial u}{\partial x} + v\frac{\partial v}{\partial y}$$

$$1 = \frac{x^2}{4} \times \frac{2x}{4} + \left(-\frac{xy}{2}\right) \times 0$$

$$\frac{x^3}{8} = 1$$

$$x = 2$$

$$\text{Also, } a_y = u\frac{\partial v}{\partial x} + v\frac{\partial v}{\partial y}$$

$$-0.5 = \frac{x^2}{4} \times \left(-\frac{y}{2}\right) + \left(-\frac{xy}{2}\right) \times \left(-\frac{x}{2}\right)$$

$$-\frac{x^2 y}{8} + \frac{x^2 y}{4} = -0.5$$

$$\frac{x^2 y}{8} = -0.5$$

$$x^2 y = -0.5$$

$$x^2 y = -4$$

$$y = -1 (\text{Since } x = 2)$$

$$\text{Therefor, } \vec{V} = \frac{2^2}{4}\hat{\imath} - \frac{2 \times (-1)}{2}\hat{j} = 1\hat{\imath} + 1\hat{j}$$

$$\therefore \left|\vec{V}\right| = \sqrt{2} = 1.4$$

Hence, the answer is 1.4 m/s.

Q5. If $\varnothing(X,Y)$ is velocity potential and $\psi(X,Y)$ is stream function for a 2-D, steady, incompressible and irrotational flow, which one of the following are incorrect?

(A) $\left(\frac{dy}{dx}\right)_{\varnothing=\text{Constant}} = -\frac{1}{\frac{dy}{dx}_{\psi=\text{Constant}}}$

(B) $\frac{\partial^2\psi}{\partial x^2}+\frac{\partial^2\psi}{\partial y^2}=0$

(C) $\left(\frac{dy}{dx}\right)_{\varnothing=\text{Constant}} = -\frac{1}{\frac{dy}{dx}_{\psi=\text{Constant}}}$

(D) $\frac{\partial^2\psi}{\partial x^2}+\frac{\partial^2\psi}{\partial y^2}=0$

Solution:

Since constant stream and constant velocity potential lines intersect each other orthogonally, therefore product of their slope at any point will be equal to -1.

Therefore, the statement (A) is correct.

For irrotational flow, the stream function satisfies the Laplace equation,

$$i.e.\ \frac{\partial^2\psi}{\partial x^2}+\frac{\partial^2\psi}{\partial y^2}=0$$

Therefore, the statement (B) is correct.

The statement (C) is incorrect since it cannot be true simultaneously with statement (A).

For incompressible flow, the velocity potential function satisfies Laplace equation,

$$i.e.\frac{\partial^2\varnothing}{\partial x^2}+\frac{\partial^2\varnothing}{\partial y^2}=0.$$

Therefore, the statement (D) is correct.

Hence, the answer is (C).

Q6. The flow field shown over a bluff body has considerably curved streamlines. A student measures pressures at points A, B, C, and D and denotes them as p_A, p_B, p_C, and p_D respectively. State which one of the following statements is true. The arrow indicates the freestream flow direction.

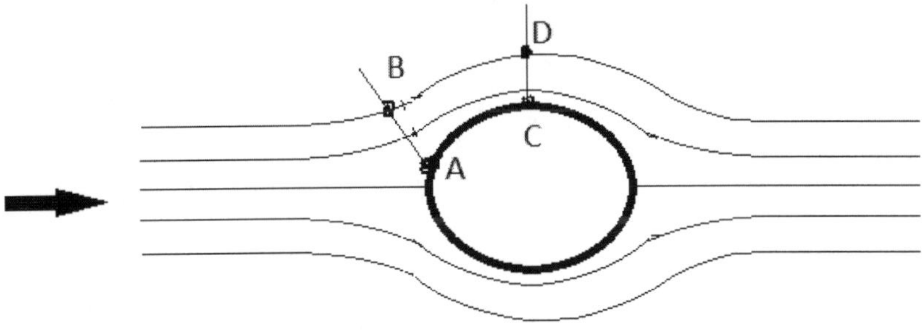

(A) $p_A = p_B$ and $p_C > p_D$

(B) $p_A > p_B$ and $p_C > p_D$

(C) $p_A = p_B$ and $p_C < p_D$

(D) $p_A > p_B$ and $p_C < p_D$

Solution:

In potential flow, at front and rear stagnation point, velocity is zero and at C is maximum.

Now, applying Bernoulli's equation at point A and B, since, $V_A < V_B$.

Therefore, $p_A > p_B$.

Now, velocity at point C is higher than D. Therefore, according to Bernoulli's theorem pressure at C should be less than that at D, i.e., $p_C < p_D$.

Hence, the answer is (D).

Q7. A 2-D incompressible flow is defined by its velocity components in m/s as $U = -\dfrac{CY}{X^2+Y^2}$ and $V = \dfrac{CX}{X^2+Y^2}$. If the value of the constant C is equal to 0.1 m²/s, the numerical value of vorticity at the point $x = 1$ m and $y = 2$ m is________ s^{-1}.

Solution:

Given, $u = -\dfrac{CY}{X^2+Y^2}$ and $v = \dfrac{CX}{X^2+Y^2}$

Vorticity is given by, $\Omega = \nabla \times V$

$$\Omega = \begin{vmatrix} \hat{i} & \hat{j} & \hat{k} \\ \dfrac{\partial}{\partial x} & \dfrac{\partial}{\partial y} & \dfrac{\partial}{\partial z} \\ u & v & w \end{vmatrix}$$

$$= \hat{i}\left(= \frac{\partial v}{\partial z} \right) - \hat{j}\left(-\frac{\partial u}{\partial z} \right) + \hat{k}\left(\frac{\partial v}{\partial x} - \frac{\partial u}{\partial y} \right)$$

$$= 0 - 0 + \hat{k}\left\{ c\frac{x^2+y^2-x(2x)}{\left(x^2+y^2\right)^2} + c\frac{x^2+y^2-y(2y)}{\left(x^2+y^2\right)^2} \right\}$$

$$= c\hat{k}\left\{ \frac{y^2-x^2+x^2-y^2}{\left(x^2+y^2\right)^2} \right\}$$

$$= 0$$

Hence, the answer is 0 s^{-1}.

Q8. Two flow configurations are shown below for flow of incompressible, viscous flow. The inlet velocity for the diverging nozzle (Fig (i)) and free-stream velocity for flow past the bluff body (Fig (ii)) is constant. Points A and B are separation points and flow is laminar. The relation regarding

velocity gradients at point A and B is (y is the direction normal to the surface at the point of separation).

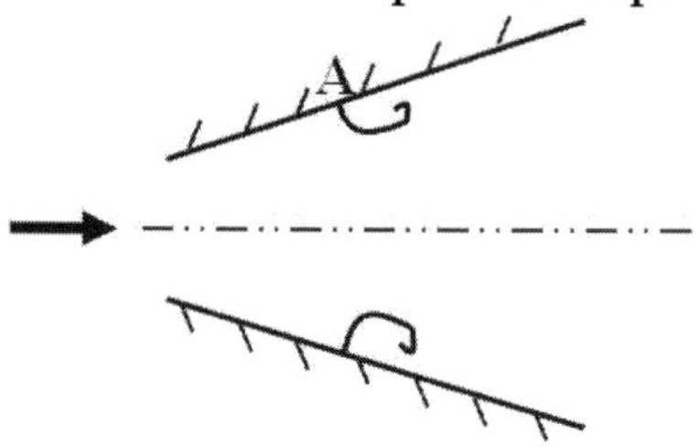

Fig (i)

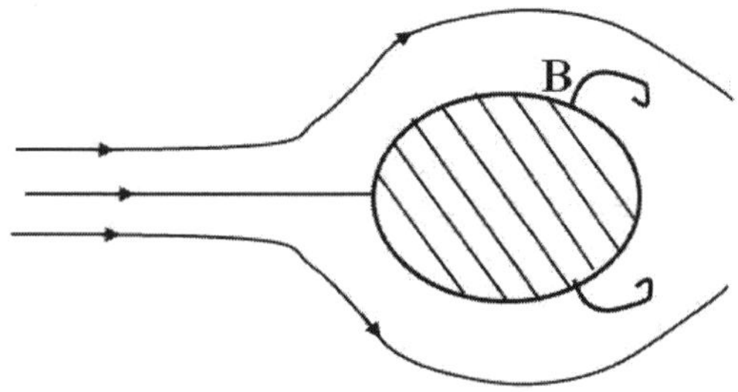

Fig (ii)

(A) $\frac{du}{dy_A} = \frac{du}{dy_B}$ (B) $\frac{du}{dy_a} > \frac{du}{dy_B}$

(C) $\frac{du}{dy_A} < \frac{du}{dy_B}$ (D) $\frac{d^2u}{dy_A^2} = \frac{d^2u}{dy_B^2}$

Solution:

At the point of separation, the wall shear stress just becomes zero, This will happen only when the velocity gradient normal to the wall becomes zero. Therefore, = .

Hence, the answer is (A).

Q9. Consider a fully developed, steady, incompressible, 2-D, viscous channel flow with uniform suction and blowing velocity V_0, as shown in the figure given below. The centerline velocity of the channel is 10 m/s along the x-direction. If the value of V_0 at both the walls is 1 m/s, the value of the y-component of velocity inside the flow field is __________ m/s.

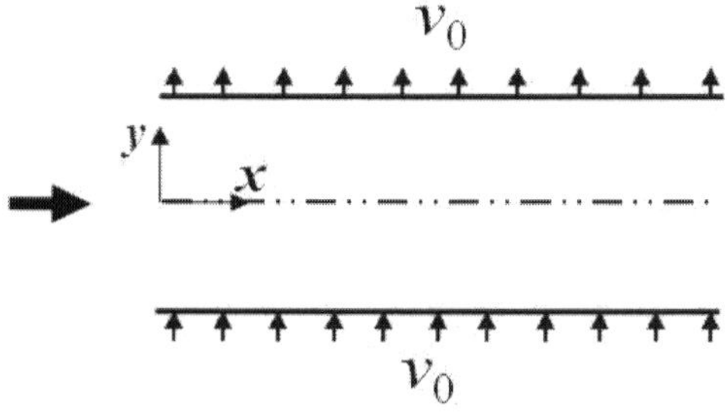

Solution:

The continuity equation for an incompressible flow is

$$\frac{\partial u}{\partial x} + \frac{\partial v}{\partial y} = 0 \quad ...(i)$$

Now, since the flow is fully developed, hence

$$\frac{\partial u}{\partial x} = 0 \qquad ...(ii)$$

Hence applying eq. (i) and eq. (ii) at any point inside the flow near the suction and blowing walls results in the following equation:

$$\frac{\partial v}{\partial y} = 0$$

$$v \neq v(y) \qquad ...(iii)$$

However, the suction and blowing velocity $V_0 = 1$ m/s.

Hence from eq. (iii), we get $v = V^0 = 1$ m/s

Hence, the answer is 1 m/s.

Q. 10 – Q. 22 carry two marks each.

Q10. Exhaust from a kitchen goes into the atmosphere through a tapered chimney as shown. The area of cross-section of chimney at location-1 is twice of that at location-2. The flow can be assumed to be inviscid with constant exhaust density of 1 kg/m^3 and acceleration due to gravity, g = 9.8 m/s^2. If the steady, uniform exhaust velocity at location-1 is U = 1m/s, the pressure drop across the chimney is ________ Pa.

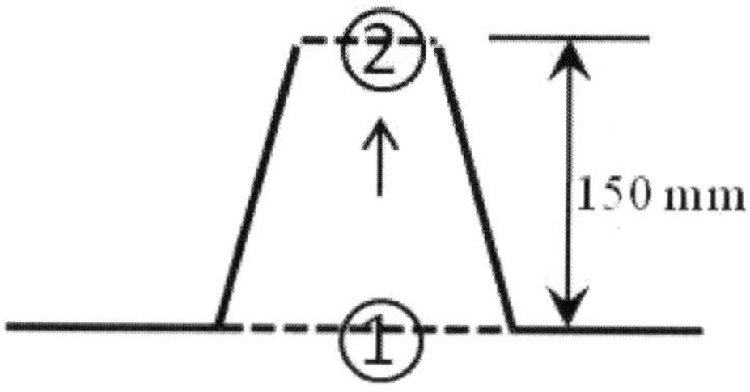

Solution:

Applying Bernoulli's equation at points 1 and 2,

$$\frac{p_1}{\rho g} + \frac{V_1^2}{2g} + z_1 = \frac{p_2}{\rho g} + \frac{V_2^2}{2g} + z_2$$

$$\frac{P_1 - P_2}{\rho g} = \frac{v_2^2 - v_1^2}{2g} + (z_2 - z_1)$$

Also, from continuity equation, we have, $A_1V_1 = A_2V_2$

$2A_2V_1 = A_2V_2$

$V_2 = 2V_1$

Given that $V_1 = 1$m/s, ρ=1 kg/m², and z_2-z_1=0.15 m

Therefore, $\frac{p_1 - p_2}{\rho g} = \frac{2V_1^2}{2g} + 0.15$

$$\frac{p_1 - p_2}{\rho g} = 0.303$$

$$p_1 - p_2 = 2.97Pa$$

Hence, the answer is 2.97 Pa.

Q11. A jet of diameter 20 mm and velocity 6 m/s coming out of a water-tank standing on a frictionless cart hits a vane and gets deflected at an angle 45° as shown in the figure below. The density of water is 1000 kg/m³. Neglect all minor and viscous losses. If the cart remains stationary, the magnitude of tension in the supporting string connected to the wall is ________ N.

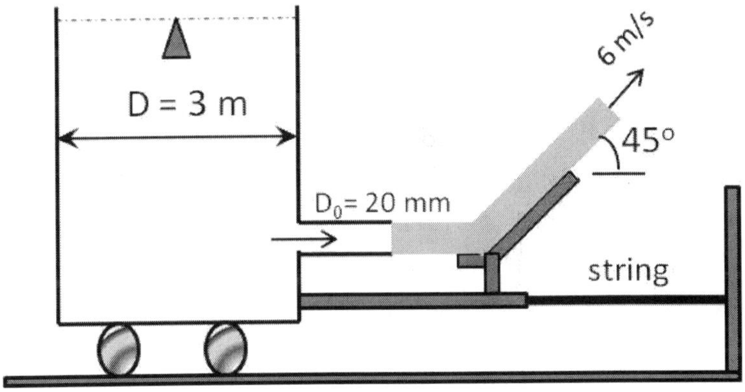

Solution:

Given, ρ = 1000kg/m³, d = 20 mm = 0.020 m, V = 6 m/s

The tension in the string connected to the wall will be equal to the net force exerted by the jet along the x-axis.

Now, force exerted by the jet in the positive x- axis,

$$F_1 = \dot{m}\{(V_x)_1 - (V_x)_2\}$$

$$= \rho AV(V - V\cos 45°)$$

$$\rho AV^2 - \rho AV^2 \cos 45°$$

$$= 1000 \times \frac{3.14 \times 0.020 \times 0.020}{4} \times 6 \times 6 \times (1 - 0.707) \text{ N}$$

3.31 N

Now, Reaction force exerted by the jet on the cart, F2 = F_2= – AV²

$$= -1000 \times \frac{3.14 \times 0.020 \times 0.020}{4} \times 6 \times 6 N$$

$$= -11.31 N$$

Therefore, total tension in the string, T = $F_2 - F_1$ = 11.31 – 3.31 N = 8 N

Hence, the answer is 8 N.

Q12. A block is floating at the oil-water interface as shown. The density of oil is two-thirds of that of water. Given that the density of the block is 800 kg/m³ and that of water is 1000 kg/m³, the fraction of the total height of block in oil is ________.

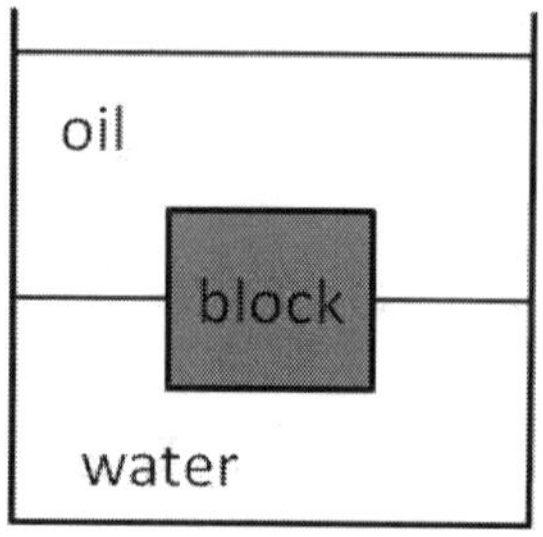

Solution:

Given:

Density of block, ρ_b = 800kg/m³

Density of water, ρ_w = 1000 kg/m³

Density of oil, ρ_o= kg/m³

Let, A = cross sectional of the block

h = Total height of the block

x = height of the block immersed in the oil.

Then, height of block immersed in the water will be $h-x$.

Now, since the block is in equilibrium, therefore, weight of the block must be equal to sum of buoyancy forces due to water plus oil.

i.e. $\rho_b \times V \times g = \rho_w \times V' \times g + \rho_o \times V'' \times g$

Where, V is the volume of block, V' is the volume of block immersed in water and V'' is the volume of block immersed in oil.

$$800 \times A \times h \times g = 1000 \times A \times (h - x) \times g + \frac{2}{3} \times 1000 \times A \times x \times g$$

$$0.8 = \left(1 - \frac{x}{h}\right) + \frac{2}{3} \times \frac{x}{h}$$

$$0.8 = 1 - \frac{1}{3} \times \frac{x}{h}$$

$$\frac{x}{h} = 0.6$$

Hence, the answer is 0.6

Q13. A horizontal pipe is feeding water into a reservoir from the top with a time-dependent volumetric flow-rate, Q (m^3/h) = 1+0.1 where t is time in hours. The area of the base of the reservoir is 0.5 m^2. Assuming that initially the reservoir was empty, the height of the water level in the reservoir after 60 minutes is _________ m.

Solution:

We have, Q = 1+ 0.1

$$i.e. \frac{d\forall}{dt} = 1 + 0.1 \times t,$$

where $\forall$ is the volume of water in the tank at any instant

Hence, $\forall - \int_0^1 (1 + 0.1t)\,dt$

$$\forall = \frac{2.1}{2} m^3$$

Thus, volume delivered through the pipe in 60 min i.e. 1 h. in the tank is $\frac{2.1}{3}$ m^3

Therefore, $\frac{2.1}{2} = A \times h,$

where A is the cross-sectional area and h is the height of water in the tank.

$$\frac{2.1}{2} = 0.5 \times h$$

$$h = 2.1 \text{ m}$$

Hence, the answer is 2.1m.

Q14. Velocity field of a 2-D steady flow is provided as $\vec{V} = c\left(x^2 - y^2\right)\hat{\imath} - 2cxy\hat{\jmath}$. The equation of the streamlines of this flow is

(A) $x^2 y - \frac{y^2}{3} = \text{Constant}$

(B) $y^2 x - \frac{y^2}{3} = \text{Constant}$

(C) $xy - \frac{y}{3} = \text{Constant}$

(D) $x^2 y - \frac{y^3}{3} = \text{Constant}$

Solution:

We have, $u = c(x^2-y^2)$ and $v = -2cxy$

$$\text{Now, } -\frac{\partial\Psi}{\partial x} = v \text{ and } \frac{\partial\Psi}{\partial y} = u$$

$$-\frac{\partial\Psi}{\partial x} = -2cxy \text{ and } \frac{\partial\Psi}{\partial y} = c\left(x^2 - y^2\right),$$

$$\Psi = 2c\int xy\,dx \text{ and } \Psi = c\int\left(x^2 - y^2\right)dy$$

$$\Psi = cx^2y + f(y) + k_1 \text{ and } \Psi = cx^2y - c\frac{y^3}{3} + k_2$$

$$\text{Therefore, } x^2y - \frac{y^3}{3} = \text{ const.}$$

Hence, the answer is (A) and (D).

Q15. Velocity potential and stream function in polar coordinates (r,θ) for a potential flow over a cylinder with radius is given as (r+)cos and = (r-) sin respectively. Here, denotes uniform freestream velocity, and θ is measured counter clockwise as shown in the figure. How does the velocity magnitude, over the surface of the cylinder ~~will~~ vary?

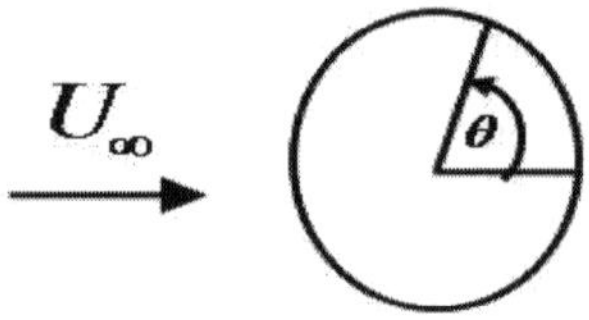

(A)cos (B)sin

(C) cos (D)sin

Solution:

$$\Psi = U_\infty\left(r - \frac{R^2}{r}\right)\sin\theta$$

$$V_\theta = \frac{d\Psi}{dr}$$

$$= U_\infty\left(1 - R^2 \times \frac{-1}{r^2}\right)\sin\theta$$

At r = R

$$V_\theta = q$$

$$= U_\infty\left(1 + \frac{R^2}{r^2}\right)\sin\theta$$

$$= 2U_\infty \sin\theta$$

Hence, the answer is (D).

Q16. Consider a laminar flow over a flat plate of length = 1 m. The boundary layer thickness at the end of the plate is δ_w for water, and δ_a for air for the same free stream velocity. If the kinematic viscosities of water and air are 1×10^{-6} m²/s and 1.6×10^{-5} m²/s, respectively, the numerical value of the ratio δ_w/δ_a, is ________.

Solution:

We know that, $\frac{\delta}{x} = \frac{5}{\sqrt{\mathrm{Re}_x}}$

$$\delta = \frac{5x}{\sqrt{\frac{U_\infty x}{\vartheta}}}$$

$$= 5\sqrt{\frac{\vartheta x}{U_\infty}}$$

Therefore for a given x-location and a given free-stream velocity

$$\delta \propto \sqrt{\vartheta}$$

Hence,

$$\frac{\delta_w}{\delta_a} = \sqrt{\frac{\vartheta_w}{\vartheta_a}}$$

$$= \sqrt{\frac{1\times10^{-6}}{1.6\times10^{-5}}}$$

$$= 0.25$$

Hence, the answer is 0.25.

Q17. Prototype of a dam spillway (a structure used for controlled release of water from the dam) has characteristic length of 20 m and characteristic velocity of 2 m/s. A small model is constructed by keeping Froude number same for dynamic similarity between the prototype and the model. What is the minimum length-scale ratio between prototype and the model such that the minimum Reynolds' number for the model is 100? The density of water is 1000 kg/ and viscosity is Pa-s.

(A) 1.8×10^{-4} (B) 1.8×10^{-4} (C) 1.8×10^{-3} (D) 9.1×10^{-4}

Solution:

We know,

$$\frac{\rho l_p}{v_p^2} = \frac{\rho l_m}{v_m^2}$$

Now, $\frac{20}{2\times2}=\frac{l_m}{v_m^2}$

$$\frac{20}{2\times2}=\frac{l_m}{v_m^2}$$

$$\frac{l_m}{v_m^2}=5 \qquad \text{...(ii)}$$

$$\text{Re}=\frac{\rho v_m l_m}{\mu}$$

$$\frac{1000\times l_m\times v_m}{10^{-3}}=100$$

$$=l_m v_m=10^{-4}$$

$$v_m=\frac{10^{-4}}{l_m} \qquad \text{...(ii)}$$

Now, putting in equation (i)

$$l_m^3=5\times10^{-8}$$

$$l_m=3.68\times10^{-3}\,m$$

$$\frac{l_m}{l_p}=\frac{3.68\times10^{-3}}{20}=1.84\times10^{-4}$$

Scale factor,

Hence, the answer is (A).

Q18. An orifice meter, having orifice diameter of $d=\frac{20}{\sqrt{\pi}}$ mm is placed in a water pipeline having flow rate $Q_{act}=3\times10^{-4}\,m^3/\text{sec}$. The ratio of orifice diameter to pipe diameter is 0.6. The contraction coefficient is also 0.6. The density of water is 1000 kg/m³. If the pressure drop across the orifice plate is 43.5 kPa, the discharge co-efficient of the orifice meter at this flow Reynolds number is________.

Solution:

Given:

Orifice meter diameter $d=\frac{20}{\sqrt{\pi}}mm$

Hence the orifice area, $a_0 = \frac{\pi}{4} d^2 = \frac{\pi}{4} \times \frac{400}{\pi} mm^2 = 10^{-4} m^2$

And, $Q_{act} = 3 \times 10^{-4} \, m^3/s$

Now, contraction coefficient, $C_C = 0.6$.

Therefore, Therefore, $\frac{a_c}{a} = 0.6$

Also, $\rho = 1000 \, kg/m^2$

And, $\nabla p = 43.5\,\text{kPa} = 43500$ Pa

$$Qlact = C_d \frac{a_0 \sqrt{2\Delta p / p}}{\sqrt{1 - c_c^2 \frac{a_o^2}{a^2}}}$$

$$c_d = \frac{3 \times 10^{-4} \times \sqrt{1 - (0.6)^4 \times (0.6)^2}}{10^{-4} \times \sqrt{2 \times 43500/1000}} = 0.31$$

Hence, the answer is 0.31

Q19. Consider the following figures shown below. The objects are marked as A1, A2, B1, B2 and C1, C2 and the flow directions over these objects are shown by the respective arrow placed to the left of the object. Freestream velocities are same for all the cases. Amongst these objects, A1, A2, B1 and C1 are having smooth surfaces while B2 and C2 are having rough surfaces. Reynolds number is such that flow over rough surfaces become turbulent and flow over smooth surfaces can be considered laminar. All the airfoils can be considered as thin slender airfoil. Among the statements (i) to (vi) made about the drag of these objects which is/are correct?

(i) Drag of object A1 is less than drag of object A2.

(ii) Drag of Object A1 and A2 are same.

(iii) Drag of Object B1 is more than drag of object B2.

(iv) Drag of object B2 is more than drag of object B1.

(v) Drag of Object C1 is more than drag of object C2.

(vi) Drag of object C2 is more than drag of object C1.

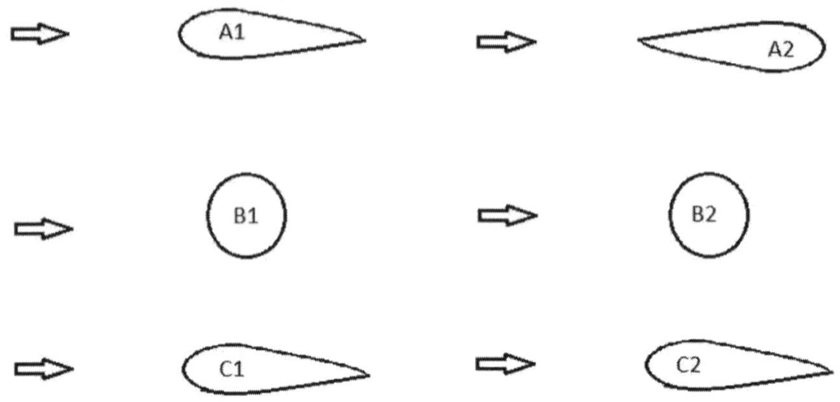

(A) (i), (iii) & (vi)
(B) (ii),(iii) & (vi)
(C) (i),(iii) & (v)
(D) (i),(iv) & (vi)

Solution:

For the smooth airfoils A1 and A2, A1 will experience less drag than A2.

For the two objects B1 and B2 having circular cross section, the one with a rough surface, resulting in turbulence will undergo delayed flow separation. The effect of delayed separation is more dominant in the reduction of form drag than the corresponding increase in skin-friction drag, resulting in an overall decrease in total drag. Therefore, the object B2 will experience less drag than B1.

In the case of airfoils C1 and C2, the ski-friction drag is more prominent than the form drag. Therefore, the one with a rough surface will experience more drag. In other words, drag of object C2 is more than that of C1.

Hence, the answer is (A).

Q20. Consider 2-D, steady, incompressible, fully developed flow of viscous, Newtonian fluid through two stationary parallel plates, in Cartesian co-ordinate (x,y,z) system. Assume plates are very long in x-direction, wide in z-direction (also there is no variation of velocity in z direction) and distance between them is $2h$. The velocity in such a channel is given as $U = U_{max}\left(1 - \frac{y^2}{h^2}\right)$. The origin y = 0 is located at the center between the plates. If h = 48 mm and $U_{max} = 100\,{}^{mm}\!/_{sec}$, difference between values of stream functions passing through y= 0 and $y = h/2$ is __________ mm²/s.

Solution:

$$U = U_{\max}\left(1 - \frac{y^2}{h^2}\right)$$

$$U = \frac{\partial \psi}{\partial y}$$

$$\frac{\partial \psi}{\partial y} = U_{\max}\left(1 - \frac{y^2}{h^2}\right)$$

$$\partial \psi = U_{\max}\left(1 - \frac{y^2}{h^2}\right)\partial y$$

$$[\psi]_1^2 = U_{\max}\left[y - \frac{y^3}{3h^2}\right]_{y=0}^{y=\frac{h}{2}}$$

$$\psi_2 - \psi_1 = U_{\max}\left\{\frac{h}{2} - \frac{h^3}{24h^2}\right\}$$

$$= U_{\max}\left(\frac{12h - h}{24}\right)$$

$$= U_{\max} \times \frac{11h}{24}$$

$$= 100 \times \frac{11 \times 48}{24}$$

$$= 2200\, m^2/s$$

Hence, the answer is 2200 m²/s.

Q.21 A pump is used to deliver water to an overhead tank at a flow rate of Q = 4 x 10^{-3} m^3/s. The pump adds 1.6 kW to water. If the density of water is 1000 kg/m3 and acceleration due to gravity is 10 m/s^2, the pump head added to the flow is ________ m.

Solution:

The power (in kW) added to a fluid by a pump is given by

$$P = \frac{\rho g Q H}{10^3},$$

where H is the head added to the flow.

$$1.6 = \frac{1000 \times 10 \times 4 \times 10^{-3} \times H}{10^3}$$

$$H = \frac{1.6\times10^3}{10\times4} = 40m$$

Hence, the answer is 40 m.

Q22. Water is discharged at atmospheric pressure from a large reservoir through a long pipe of diameter *d* and length *L*. The height of the free surface of the reservoir from the discharge point is *h* meters. The Darcy's friction factor of the pipe is 0.002. Neglect the velocity inside the reservoir as the reservoir is very large. Given, $L = 20$ m, $d = 40$ mm, density of water = 1000 kg/m^3 and flow rate is $Q = 4\pi \times 10^{-3}$ m^3/s. Assume gravitational acceleration, $g = 10$ m/s^2. The value of *h* is _______ m.

Solution:

Writing the energy balance between the outlet of the tank and the outlet of the pipe we get

$$h = \frac{V^2}{2g} + h_f$$

$$= \frac{\left(\frac{4q}{\pi D^2}\right)^2}{2g} + \frac{fLV^2}{2gD}$$

$$= \frac{16Q^2}{2g\pi^2 D^4} + \frac{fL\times\left(16Q^2/\pi^2 D^4\right)}{2gD}$$

$$= \frac{16\times\left(4\pi\times10^{-3}\right)^2}{2g\pi^2 D^4} + \frac{16\times fL\times\left(4\pi\times10^{-3}\right)^2}{2g\pi^2 D^5}$$

$$= \frac{16\times16\pi^2\times10^{-6}}{2\times10\times\pi^2\times(0.04)^4} + \frac{8\times0.002\times20\times16\pi^2\times10^{-6}}{10\times\pi^2\times(0.04)^5}m$$

$$= 10 \text{ m}$$

Hence, the answer is 10 m.

GATE 2015
FLUID MECHANICS

Q. 1 – Q. 9 carry one mark each.

Q1. **The gap δ between two concentric cylinders, each of height *h*, is filled with an oil. The torque required to rotate the inner cylinder at an angular velocity of ω against the fixed outer cylinder is *T*. The diameter of the inner cylinder is *d* and δ<<d. The dynamic viscosity of the oil is given by**

(A) $\dfrac{4\pi\delta T}{d^3\omega h}$ (B) $\dfrac{4\delta T}{\pi d^3\omega h}$

(C) $\dfrac{4\pi\delta T}{d^2\omega h^2}$ (D) $\dfrac{4\delta T}{\pi d\omega h^3}$

Solution:

Given, τ = T and δ ≪ d

Now, the torque required to rotate the inner cylinder is given by, $T = F \times \frac{d}{2}$ ... (1)

Also, we have $\tau = \frac{F}{A}$ ... (2)

where τ is the shear stress.

Now, from (1) we have, $F = \frac{2T}{d}$ ...(3)

Putting value of F from eq. (3) in eq. (2), we get, = $\tau = \frac{\frac{2T}{d}}{\pi d h}$

$$\tau = \frac{2T}{\pi d^2 h}$$

Now, from Newton's law of viscosity, we have,

$$\tau = \mu \frac{du}{dy}$$

Therefore, = $\mu = \frac{\tau}{\frac{du}{dy}}$

$$\mu = \frac{\frac{2T}{\pi d^2 h}}{\frac{\omega d}{2\delta}} \left(\because \frac{du}{dy} = \frac{U}{\delta} = \frac{\omega d}{2\delta} \right)$$

$$\mu = \frac{(2T)(2\delta)}{\pi d^3 h\omega}$$

$$\mu = \frac{4\delta T}{\pi d^3 \omega h}$$

Hence, the answer is (B).

Q2. Water is retained against a sluice gate in the form of a circular segment as shown in the figure. If ϱ and g are the density of water and gravitational acceleration respectively, the upward force exerted by the gate on the water per unit depth perpendicular to the plane of the figure is

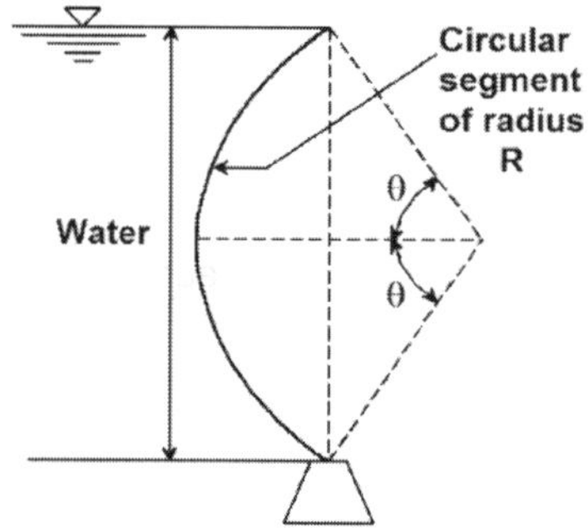

(A) $QR^2(\theta - \frac{1}{2}\sin 2\theta)g$ (B) $QR^2(\cos^2\theta - \frac{1}{2}\sin 2\theta)g$

(C) $QR^2(\cos\theta - \frac{1}{2}\sin 2\theta)g$ (D) $QR^2(\cos^2\theta - \frac{1}{2}\sin^2\theta)g$

Solution:

Upward force exerted by the fluid is equal to weight of the fluid supported by the curved surface.

i.e. $F_y = \rho \forall g$

where is the volume of the fluid supported by the curved surface.

Now, ∀=AL

where A is the cross-sectional area of the curved surface and L is the length of curved surface perpendicular to the plane of paper.

Therefore, $\forall = R^2\theta - R\sin\theta \times R\cos\theta$ ($\because L = 1$)

$$= R^2\left(\theta - \frac{1}{2} \times 2\sin\theta\cos\theta\right)$$

$$= R^2\left(\theta - \frac{1}{2}\sin 2\theta\right)$$

Therefore, $F_y = \rho R^2 \left(\theta - \frac{1}{2}\sin 2\theta \right) g$

Hence, the answer is (A).

Q3. A two-dimensional velocity field is given by $\vec{V} = 10\left(y^3 - x^2 y\right)\hat{\imath} + 2Cxy^2 j$, where and are the unit vectors in the dimensions of the rectangular Cartesian coordinates *x* and *y*, respectively. If the flow is incompressible, the constant C should be

(A) -10 (B) 0

(C) 5 (D) 10

Solution:

Given, $\vec{V} = 10\left(y^3 - x^2 y\right)\hat{\imath} + 2Cxy^2 j$

Here, and u = $10(y^3 - x^2 y)$ and $v = 2Cxy^2$

Therefore, and $\frac{\partial u}{\partial x} = -20xy$ and $\frac{\partial v}{\partial y} = 4Cxy$

Since, flow is incompressible, the continuity equation must be satisfied, i.e.,

$$\frac{\partial u}{\partial x} + \frac{\partial v}{\partial y} = 0$$

$-20xy + 4Cxy = 0$

$4xy(C-5) = 0$

C = 5

Hence, the answer is (C).

Q4. Let $\vec{V}$ and T denote the velocity vector and temperature in a flow field. The rate of change of temperature experienced by a fluid particle as it is passes through the point (x_1, y_1, z_1) at a time t_1 is 2.5 °C/s. The rate of change of temperature at a time t_1 at the point (x_1, y_1, z_1) is 4.8 °C/s. The quantity $\vec{V}.\nabla T$ at (x_1, y_1, z_1) and t_1 in °C/s is _____.

Solution:

Here, $T = f(x,y,z,t)$

Therefore, $\frac{DT}{Dt} = \frac{\partial T}{\partial x}\frac{dx}{dx} + \frac{\partial T}{\partial y}\frac{\partial T}{\partial y} + \frac{\partial T}{\partial z}\frac{dz}{dt} + \frac{\partial T}{\partial t}$

$$\frac{DT}{Dt} = u\frac{\partial T}{\partial x} + v\frac{\partial T}{\partial y} + w\frac{\partial T}{\partial z} + \frac{\partial T}{\partial t}$$

$$\frac{DT}{Dt} = \vec{V}.\nabla T + \frac{\partial T}{\partial t}$$

$$\vec{V}.\nabla T = \frac{DT}{Dt} - \frac{\partial T}{\partial t} = 2.5 - 4.8 = -2.3°\text{C/s}$$

Hence, the answer is -2.3°C/s.

Q5. In a simple Couette flow apparatus, the gap h between the parallel plates is filled with a liquid of density ϱ and dynamic viscosity μ, and one plate is dragged at a velocity of U parallel to itself, while the other plate is fixed. The magnitude of vorticity at any point in the flow is

(A) $\frac{\mu}{\rho h^2}$ (B) 0

(C) $\frac{1}{h^2}\sqrt{\frac{\mu U h}{\rho}}$ (D) $\frac{U}{H}$

Solution:

Vorticity is given by, $\Omega = \frac{\partial v}{\partial x} - \frac{\partial u}{\partial y}$

Now, for a simple Couette flow velocity profile is given by,

$\frac{u}{U} = \frac{y}{H}$, or $u = \frac{y}{H}$ and $v = 0$.

Therefore, $\Omega = \frac{\partial v}{\partial x} - \frac{\partial u}{\partial y} = 0 - \frac{\partial\left(U\frac{y}{H}\right)}{\partial y} = -\frac{u}{H}$.

Therefore, the magnitude of vorticity is $\frac{U}{H}$

Hence, the answer is (D).

Q6. An open glass capillary tube of 2 mm bore is lowered into a cistern containing mercury (density = 13600 kg/m³) as shown in the figure. Given that the contact angle between mercury and glass = 140°, surface tension coefficient = 0.484 N/m and gravitational acceleration = 9.81 m/s², the depression of mercury in the capillary tube below the free surface in the cistern, in mm, is _____.

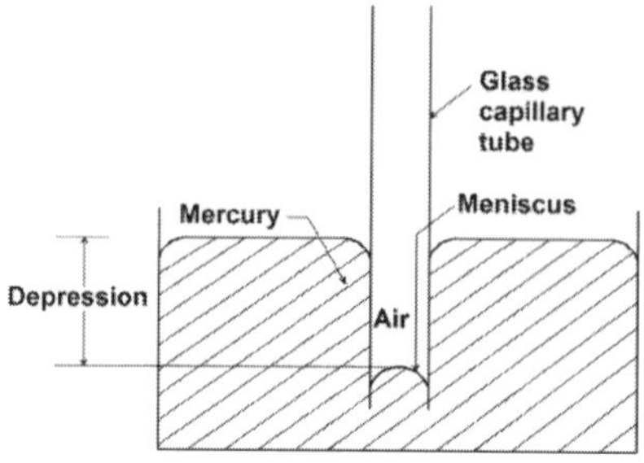

Solution:

Given:

Diameter, $d = 2$ mm = 0.002 m,

Density, kg/m^3

Angle of contact, $\theta = 140^0$

Surface tension, $\sigma = 0.484$ N/m

Now, the depression of mercury in the capillary tube is given by, $h = \dfrac{4\sigma\cos\theta}{\rho g d}$

Therefore, $h = \dfrac{4\times0.484\times\cos140}{13600\times9.8\times0.002} = \dfrac{-1.483}{266.56} = 0.0055\text{ m} = 5.5\text{ mm}$

Hence, the answer is 5.5 mm.

Q7. Consider a combined forced-free vortex. The central region with radius R and angular velocity ω is the forced vortex and the rest is the free vortex. The pressure at the edge of the combined vortex is. If the density of the fluid is ρ, the pressure at the center of the combined vortex is

(A) $\rho\omega^2R^2$ (B) ½$\rho\omega^2R^2$ (C) ½$\rho\omega^2R^2$ (D) + $\rho\omega^2R^2$

Solution:

For forced vortex flow, pressure distribution is given as

$$p_1 - p_2 = \frac{1}{2}\rho\omega^2\left(r_1^2 - r_2^2\right)$$

Where, are the pressures at radial distance r_1 and r_2, respectively.

Now, if p_1 is the pressure at the Centre of combined flow and p_2 is the pressure at distance R

Then, $p_1 - p_2 = \dfrac{1}{2}\rho\omega^2\left(o - R^2\right)$

$$p_1 - p_2 = \frac{1}{2}\rho\omega^2 R^2 \qquad \ldots(1)$$

Now, Applying Bernoulli's theorem in free vortex flow region.

$$p_1 - p_2 = \frac{1}{2}\rho\omega^2 R^2 = p_0 \qquad \ldots(2)$$

After substituting for p_1 from eq. (1) in eq. (2)

$$\left(p_1 + \frac{1}{2}\rho\omega^2 R^2\right) + \frac{1}{2}\rho\omega^2 R^2 = p_0$$

$$p_1 = p_0 - \rho\omega^2 R^2$$

Hence, the answer is (A).

Q8. The following is true at the point of separation of a boundary layer,

(A) Transition occurs from laminar to turbulent flow

(B) The flow relaminarizes from turbulent regime

(C) The shear stress vanishes

(D) The relation between stress and rate of strain ceases to be linear

Solution:

At the point of separation of boundary layer, the velocity gradient at the wall vanishes and therefore the shear stress is zero.

Hence, the answer is (C).

Q9. A certain fluid flow phenomenon is described by Reynolds, Weber and Ohnesorge numbers. The Weber and Ohnesorge numbers are defined as We = $\frac{\rho U^2 L}{\sigma}$, and Oh $= \frac{\mu}{\sqrt{\rho \sigma L}}$, respectively, where σ is the surface tension, ρ is the density, μ is the dynamic viscosity, U is the velocity and L is the characteristic dimension. If Re denotes the Reynolds number, which of the following relations is true?

(A) We= Oh Re^2

(B) We= Oh^2 Re^2

(C) We= Oh^2 Re

(D) We= Oh Re

Solution:

We have, Weber number, We $= \frac{\rho U^2 L}{\sigma}$ and Ohnesorge number, Oh $= \frac{\mu}{\sqrt{\rho \sigma L}}$

Also, we have Reynold number, $\text{Re} = \frac{\rho U L}{\mu}$

Now, $Oh^2 \times Re^2$

$$= \frac{\mu^2}{\rho \sigma L} \times \frac{\rho^2 U^2 L^2}{\mu^2}$$

$$= \frac{\rho U^2 L}{\sigma}$$

$$= \text{We}$$

Hence, the answer is (B).

Q. 10 – Q. 22 carry two marks each.

Q10. A rectangular boat 6 m wide and 15 m long (dimension perpendicular to the plane of the figure) has a draught of 2 m. The side view of the boat is as shown in the figure. The centre of gravity G of the boat is the free surface level. The metacentric height of the boat in m is

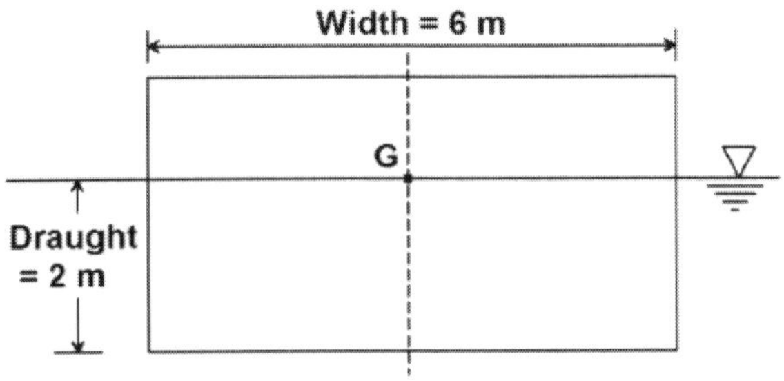

(A) −1.0 (B) 0.5 (C) 1.5 (D) 2.0

Solution:

Given:

Width of the boat, d = 6 m

Length of the boat, b = 15 m

Now, metacentric height is given by, GM $= \frac{I_G}{\forall} - BG$

where, I_G = second moment of area of plane of the boat about longitudinal axis passing through centroid of the rectangle.

BG = Distance between Centre of gravity and Centre of buoyancy.

And, ∀ = Volume of the portion of boat immersed in the water.

Now, $I_G = \frac{1}{12} bd^3$

Therefore, $I_G = \frac{1}{12} \times 15 \times 6^3$ and $\forall = 2 \times 6 \times 15\ \text{m}^3$

Now, distance of Centre of gravity from bottom, AG = 2 m

Distance of Centre of buoyancy from bottom, AB = $\frac{2}{2} = 1$ m.

Therefore, BG= 2-1 = 1 m.

$$\text{Therefore, GM} = \frac{15 \times 6^3}{12 \times 15 \times 2 \times 6} - 1$$

$$= \frac{6 \times 6}{12 \times 2} - 1$$

$$= 0.5$$

Hence, the answer is (B).

Q11. Water drains out into atmosphere from a small orifice located at the bottom of a large open tank. If the initial height of the water columns is *H*, the time taken to empty the tank is proportional to

(A) $H^{1/2}$ (B) H (C) $H^{3/2}$ (D) H^2

Solution:

Given, initial height of water column = H.

Let, at any time t, the height of the water level be h and in small time interval dt, the water level decreases by dh.

Let A be the cross-sectional area of the tank and a be cross sectional area of the orifice.

Now, flow velocity at the orifice after time t is given by $V = \sqrt{2gh}.$

Therefore, volumetric flow rate of the orifice is given by, $Q = a \times V = a \times \sqrt{2gh}$

Also, the volumetric flow rate of the tank is given by, $Q' = A \times -\frac{dh}{dt}$

Now, from continuity equation: $Q = Q'$

$$\text{Therefore, } a \times \sqrt{2gh} = -A \times \frac{dh}{dt}$$

$$dt = -\frac{A}{a}\frac{dh}{\sqrt{2gh}}$$

$$t = \frac{A}{a\sqrt{2g}}\int_H^0 \frac{-dh}{\sqrt{h}}$$

$$t = \frac{A}{a\sqrt{2g}}H^{\frac{1}{2}}$$

$$t \propto H^{\frac{1}{2}}$$

Hence, the answer is (A).

Q12. Consider a two-dimensional velocity field given by $\vec{V} = \pi y\hat{\imath} - \pi x\hat{\jmath}$, where $\hat{\imath}$ and $\hat{\jmath}$ are the unit vectors in the in the directions of the rectangular Cartesian coordinates and, respectively. A fluid particle is located initially at the point (-1, 1). Its position after unit time is

(A) (−2,−2) (B) (1,−1) (C) (1,1) (D) (3,−1)

Solution:

Now, and

The velocity field is given as

$$\vec{V} = \pi y\hat{\imath} - \pi x\hat{\jmath} \quad \ldots\text{(i)}$$

From eq. (i), it is evident that the flow is steady, since there is no time-dependent term in the velocity field. Therefore, the pathline and streamline are identical. Hence the equation of the streamline can be used to predict the trajectory of a fluid particle.

From eq. (i), $u = \pi y$ and $v = -\pi x$. Now, the equation of the streamline is written as

$$\frac{v}{u} = \frac{dy}{dx} \qquad \text{...(ii)}$$

$$\frac{dy}{dx} = -\frac{x}{y}$$

$$xdy = -ydy$$

The integration of eq. (ii) gives

$$x^2 + y^2 = c \qquad \text{...(iii)}$$

The fluid particle was at (-1, 1) at time t=0. Hence from eq. (iii) we get c=2.
Hence the equation of streamline is

$$x^2 + y^2 = 2 \qquad \text{...(iv)}$$

From eq. (iv) we can express y in terms of x as

$$x = \sqrt{2 - y^2} \qquad \text{...(v)}$$

Now, $v = -\pi x$

$$\frac{dy}{dt} = -\pi\sqrt{2 - y^2}$$

$$\int_{y=1}^{y} \frac{dy}{\sqrt{2 - y^2}} = -\pi\int_{t=0}^{t=1} dt$$

$$\left[\sin^{-1}\left(\frac{y}{\sqrt{2}}\right)\right]_{y=1}^{y} = -\pi\left[t\right]_{t=0}^{t=1} \qquad \text{...(vi)}$$

Solving eq. (vi) gives $y = -1$. Similarly, the expression for the x-velocity component can be used to find the x-coordinate of the particle after unit time and it can be found as x=1.

Hence, the answer is (B).

Q13. The figure shows a reducing area conduit carrying water. The pressure p and velocity V are uniform across sections 1 and 2. The density of water is 1000 kg/m³. If the total loss of head due to friction is just equal to the loss of potential head between the inlet and the outlet, then V_2 in m/s will be _____

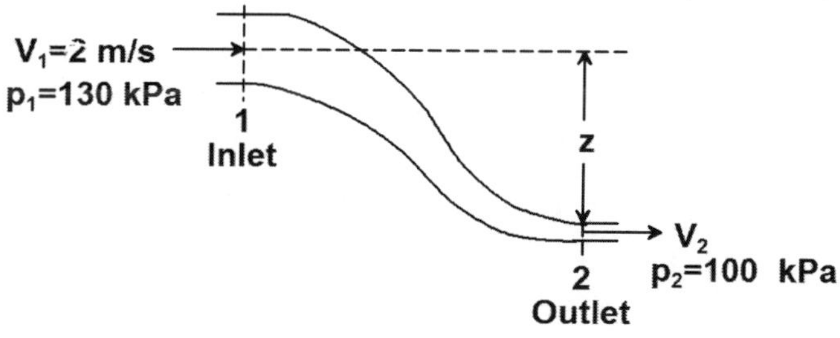

Solution:

Applying Bernoulli's theorem between points 1 and 2, we have,

$$\frac{p_1}{\rho g}+\frac{v_1^2}{2g}+z_1=\frac{p_2}{\rho g}+\frac{v_2^2}{2g}+z_2+h_f$$

$$z_1-z_2=\frac{P_2}{\rho g}+\frac{v_2^2}{2g}-\frac{P_1}{\rho g}-\frac{v_1^2}{2g}+h_f$$

$$\frac{103\times10^3}{1000g}-\frac{100\times10^3}{1000g}=\frac{v_2^2}{2g}-\frac{2^2}{2g}$$

$$30=v_2^2-4$$

$$60+4=v_2^2$$

$$64=v_2^2$$

$$v_2=8\,\text{m/s}$$

Hence, the answer is 8 m/s.

Q14. A fluid enters a control volume through an inlet port (denoted with subscript 'i') and leaves through two outlet ports (denoted with subscripts 'o,1' and 'o,2') as shown in the figure. The velocities may be assumed to be uniform across the ports. The rate of change of mass in the control volume in kg/s, at the instant shown in the figure, is _____

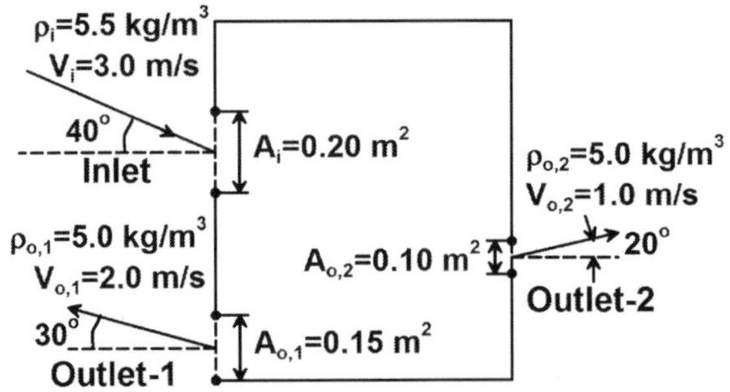

Solution:

From the conservation of mass principle, the rate of change of mass in the control volume is given by

$$\frac{dm}{dt}=\rho_1A_1\left(V_n\right)_1-\rho_2A_2\left(V_n\right)_2-\rho_3A_3\left(V_n\right)_3,$$

where is the velocity component normal to the inlet or outlet surface.

$$\therefore\frac{dm}{dt}=5.5\times0.20\times3.0\cos40°-5.0\times0.10\times10\cos20°-5.0\times0.15\times2.0\cos30°=0.75$$

Hence, the answer is 0.75.

Q15. The total discharge of water through a lawn sprinkler shown in the figure is one liter per minute. The velocity of the jet at each end, relative to the arm, is 2m/s. The density of water is 1000 kg/m³ and the length of each arm is 0.1 m. If the frictional torque of the pivot is mN.m, the rotational speed, in revolutions per minute, of the sprinkler is _____

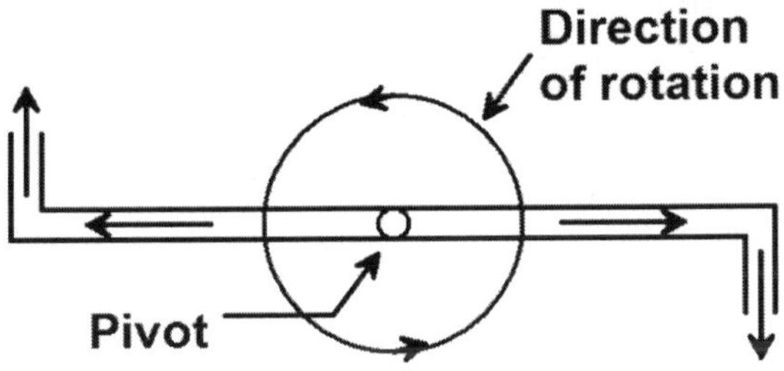

Solution:

Given:

Total discharge of water, $Q = 1$ L/min $= \frac{10^{-3}}{60}$ m/s.

Velocity of the jet relative to the arm, $V_r = \frac{2\pi}{30}$ m/s

Density of water, ρ =1000 kg/s

Length of each arm, L = 0.1m.

Friction torque, $T_f = \pi/36$ mNm.

Total kinetic energy generated per second by sprinkler, $E = \frac{1}{2}\rho Q V_r^2$.

Now, from the law of conservation of energy, total generated K.E. must be equal to sum of energy dissipated due to friction and K.E generated by rotation of sprinkle.

$$i.e. \frac{1}{2}\rho Q V_r^2 = T_f \times \omega + \frac{1}{2}\rho Q V^2$$

$$\frac{1}{2}\times 1000 \times \frac{10^{-3}}{60} \times \left(\frac{2\pi}{30}\right)^2 = \frac{\pi}{36}\times 10^{-3}\omega + \frac{1}{2}\times 1000 \times \frac{10^{-3}}{60}\times(0.1\times\omega)^2 \; (\therefore V = \omega \times L)$$

$$3.655\times10^{-4} = 8.727\times10^{-5}\times\omega + 83.333\times10^{-6}\times\omega^2$$

$$\omega = 1.64 \text{ red/s}$$

If N is the rotational speed in revolutions per minute, then

$$N = \frac{60\omega}{2\pi}$$

Solving above equation, we get, $N = 15.66$ rpm

Hence, answer is rpm.

Q16. Consider a two-dimensional potential flow field with the radial and tangential velocity components, $v_r = \frac{m}{2\pi r}$ and $v_\theta = \frac{k}{2\pi r}$, respectively, where and are constants. The stream function is such that it increases along the direction of traverse of a line in the flow field if the flow is from left to right across that line. The stream function ψ for this flow field, with $\psi = 0$ at and, $\theta = 0$ is

(A) $\frac{m\pi\theta}{2} + \frac{k\pi}{2}\ln\frac{r}{a}$ (B) $\frac{m\pi\theta}{2} - \frac{k\pi}{2}\ln\frac{r}{a}$

(C) $\frac{m_\theta}{2\pi} + \frac{k}{2\pi}\ln\frac{r}{a}$ (D) $\frac{m\theta}{2\pi} - \frac{k}{2\pi}\ln\frac{r}{a}$

Solution:

Given: $v_r = \frac{m}{2\pi r}$ and $v_\theta = \frac{k}{2\pi r}$

Now, we have

$$v_r = \frac{1}{r}\frac{d\psi}{d\theta} \quad \text{and} \quad v_\theta = -\frac{\partial \psi}{\partial r}$$

$$\frac{m}{2\pi r} = \frac{1}{r}\frac{d\psi}{d\theta} \quad \text{and} \quad \frac{-k}{2\pi r} = \frac{d\psi}{dr}$$

$$\frac{d\psi}{d\theta} = \frac{m}{2\pi} \quad \cdots(1) \quad \text{and} \quad \frac{d\psi}{dr} = -\frac{-k}{2\pi r} \quad \cdots(2)$$

Now, Integrating (2) we get,

$$\int d\psi = \int -\frac{k}{2\pi r}dr$$

$$\text{or, } \psi = -\frac{k}{2\pi}\ln(r) + c_1 \quad \ldots(3)$$

It is given that at $\psi = 0$ at $r = a$ ...(4)

Hence substituting eq. (4) in eq. (3) we get $c_1 = \frac{k}{2\pi}\ln(a)$ ...(5)

Therefore eq. (3) can be written as $\psi = -\frac{k}{2\pi}\ln\left(\frac{r}{a}\right)$ ...(6)

Again, integrating eq. (1) gives

$$\int d\psi = \int \frac{m}{2\pi} d\theta$$

$$\psi = \frac{m\theta}{2\pi} + c_2 \qquad ...(7)$$

It is also given that at θ = 0. ...(8)

Hence by substituting eq. (8) in eq. (7) we get

$$c_2 = 0 \qquad ....(9)$$

Therefore, eq. (7) can be written as

$$\psi = \frac{m\theta}{2\pi} \qquad ...(10)$$

By combining eq. (6) and eq. (8) we get

Therefore, $\psi = \frac{m\theta}{2\pi} - \frac{R}{2\pi} \ln \frac{r}{a}$

Hence, the answer is (D).

Q17. A steady, two-dimensional, inviscid and incompressible flow field is described in rectangular Cartesian coordinates as and , where and are the components of the velocity vector in the and directions, respectively. Gravity acts along the negative -direction. The pressure distribution, with the reference pressure taken as zero at the origin, with usual notation, is given by

(A) $-\frac{1}{2}\rho a^2 \left(x^2 + xy + y^2\right) - \rho gy$ (B) $-\frac{1}{2}\rho a^2 \left(x^2 + xy + y^2\right) - \rho gy$

(C) $-\frac{1}{2}\rho a^2 \left(x^2 + y^2\right) - \rho gy$ (D) $-\frac{1}{2}\rho a^2 \left(x^2 + y^2\right) - \rho gy$

Solution:

Given, $u = ax$ and $v = -$ ay

From Euler equation, we have, $u\frac{\partial u}{\partial x} + v\frac{\partial u}{\partial y} = F_x - \frac{1}{\rho}\frac{\partial p}{\partial x}$

where, F_x is the body force in x direction.

Therefore, $ax \times a + 0 = 0 - \frac{1}{\rho}\frac{\partial p}{\partial x}(F_x = 0)$

$$a^2 x = -\frac{1}{\rho}\frac{\partial p}{\partial x}$$

$$\int dp = -\rho a^2 \int x dx + f_1(y)$$

$$p = -pa^2 \int_0^x x dx + f_1(y)$$

$$p = -pa^2 \frac{x^2}{2} + f_1(y) \quad \text{...(i)}$$

Now, $u\frac{\partial v}{\partial x} + v\frac{\partial v}{\partial y} = F_y - \frac{1}{\rho}\frac{\partial p}{\partial y}$

$$ax \times 0 + (-ay \times -a) = -g - \frac{1}{\rho}\frac{\partial p}{\partial y}$$

$$a^2 y = -g - \frac{1}{\rho}\frac{\partial p}{\partial y}$$

$$\int dp = -\rho a^2 \int y dy - \int \rho g dy + f_2(x)$$

$$p = -\rho g y - \rho a^2 \frac{y^2}{2} + f_2(x) \quad \text{...(ii)}$$

Hence from eq. (i) and eq. (ii)

$$p = -\frac{1}{2}\rho a^2 (x^2 + y^2) = \rho g y$$

Hence, the answer is (C).

Q18. A cylinder of radius 0.1 m rotating clockwise about its own axis at an angular velocity of 100/ radians per second is placed in a cross-stream of air flowing at a velocity of 10 m/s from left to right. The density of air is 1.2 kg/m³. The lift force per unit length of the cylinder in N/m is _____

Solution:

Given:

Radius of cylinder, $R = 0.1$ m

Angular velocity, $\omega = \frac{100}{\pi}$ = radian/sec

Velocity of air, $V = 10$ m/s

Density of air, $\rho = 1.2$ kg/m^3

Lift force per unit length is given by, $F_L = \rho V \Gamma$

Where, Γ is the circulation of flow.

Now, circulation is given by, $\Gamma = 2\pi R^2$

Therefore, $\Gamma = 2$ m^2/s $= 2$ m^2/s

Therefore, $F_L = 1.2 \times 10 \times 2 = 24$ N/m

Hence, the answer is 24 N/m.

Q19. The velocity profile in turbulent flow through a pipe is approximated $\frac{u}{u_{\max}} = \left(\frac{y}{R}\right)^{1/7}$ as where is the maximum velocity, R is the radius and is the distance measured normal to the pipe wall towards the centerine. If denotes the velocity, the ratio $\frac{u_{av}}{u_{\max}}$ is

(A) $\frac{2}{15}$ (B) $\frac{1}{5}$ (C) $\frac{1}{3}$ (D) $\frac{49}{60}$

Solution:

Considering r as the radial distance from the centerline (refer Fig. 1), the average velocity can be obtained by

$$u_{av} = \frac{1}{\pi R^2}\int_0^R u \times 2\pi r dr$$

$$= \frac{2}{R^2}\int_0^R u_{\max}\left(\frac{R-r}{R}\right)^{1/7} rdr\left[\because y = R-r\right]$$

After carrying out the above integration by parts we get

$$u_{ac} = \frac{49}{60} \times u_{\max}$$

$$\frac{u_{av}}{u_{\max}} = \frac{49}{60}$$

Hence, the answer is (D).

Q20. A steel sphere (density = 7900 kg/m^3 of diameter 0.1 m is dropped from rest in water (density=1000 kg/m^3). The gravitational acceleration is 9.81 m/s^2. Assuming that the drag coefficient is constant and equal to 1.33, the terminal velocity attained by the sphere in m/s is _____

Solution:

Given:

Density of sphere, = 7900 kg/m^3

Diameter of sphere, D = 0.1 m

Density of water, ρ = 1000 kg/m^3

Drag coefficient, C_D = 1.33

The forces acting on the sphere are:

Weight of sphere, $W = \rho_s$ Vg

W = 7900 N = 40.55N

$$W = 7900 \times \frac{4}{3}\pi\left(\frac{0.1}{2}\right)^3 \times 9.81\text{ N} = 40.55\text{N}$$

Buoyance force,

$$F_B = \rho_w \times V \times g = 1000 \times \frac{4}{3}\pi \times \left(\frac{0.1}{2}\right)^3 \times 9.81\text{ N} = 5.133\text{ N}$$

And drag force,

$$F_D = C_D \times \frac{1}{2} \times \rho_w A V^2$$

$$= 1.33 \times \frac{1}{2} \times 1000 \times \frac{\Pi}{4} \times (0.1)^2 \times V = 5.223V^2$$

Now, when terminal velocity is achieved, net force must be zero.

i.e. $W = F_D + F_B$

$40.55 = 5.223V^2 + 5.133$

$V^2 = 6.78$

$V = 2.6$ m/s

Hence, the answer is 2.6 m/s.

Q21. An inclined venturimeter connected to an inverted manometer as shown in the figure. The cross sectional areas at the inlet and the throat are 2×10^{-3} m^2 and 2×10^{-4} m^2, respectively. The densities of water and oil are 1000 kg/m^3 and 800 kg/m^3, respectively. The gravitational acceleration is 9.8 m/s^2. If the discharge of water through the venturimeter is 5×10^{-4} m^3/s, neglecting viscous effects and assuming uniform velocities across the inlet and the throat, the manometer reading *h*, in m, will be _______.

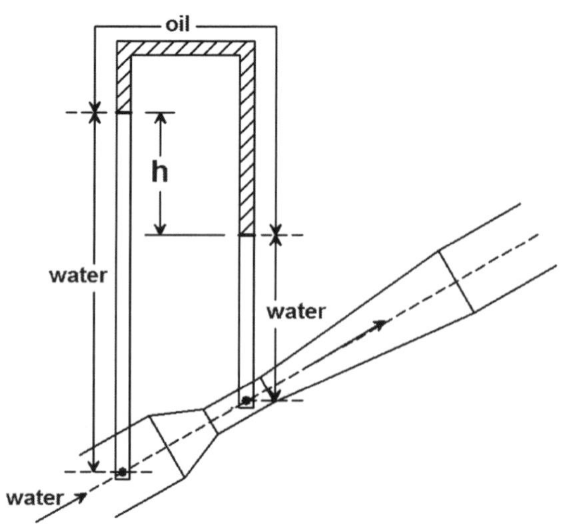

Solution:

Applying Bernoulli's theorem between point 1 and 2, we have,

$$\frac{p_1}{\rho g}+\frac{v_1^2}{2g}+z_1=\frac{p_2}{\rho g}+\frac{v_2^2}{2g}+z_2$$

$$\left(\frac{p_1}{\rho g}+z_1\right)-\left(\frac{p_2}{\rho g}+z_2\right)=\frac{v_2^2}{2g}-\frac{v_1^2}{2g}$$

$$h^*=\frac{v_2^2}{2g}-\frac{v_1^2}{2g},$$

where h^* is the piezometric head difference.

$$v_1=\frac{Q}{a_1} \quad \text{and} \quad v_2=\frac{Q}{a_2}$$

$$\frac{5\times10^{-4}}{2\times10^{-3}}=v_1 \quad \text{and} \quad \frac{5\times10^{-4}}{2\times10^{-4}}=v_2$$

$$0.25=v_1 \quad \text{and} \quad 2.5=v_2$$

$$\text{Therefore, } h=\frac{v_2^2}{2g}-\frac{v_1^2}{2g}=\frac{2.5^2}{2g}-\frac{0.25^2}{2g}$$

$$h=\frac{6.25-0.0025}{2\times9.8}=0.315\text{ m}$$

$$\text{Also, we have, } h^*=h\left(1-\frac{\rho_0}{\rho_w}\right)$$

$$0.315=h\left(1-\frac{800}{1000}\right)$$

$$\frac{0.315}{0.2}=h$$

$$x=1.575\text{ m}$$

Hence, the answer is 1.575 m.

Q22. A plane jet of water with volumetric flow rate 0.012 m³/s and cross-sectional area 6 x 10⁻⁴ m² strikes a stationary plate inclined at angle θ and leaves as two streams, as shown in the figure. The ratio of the discharge through section 2 to that through section 3 is 3:1. The velocities may be considered uniform across the cross-sections and the effects of friction may be neglected. The density of water is 1000 kg/m³. Ignoring the effects of gravity, the magnitude of the normal force exerted on the plate, in N, is

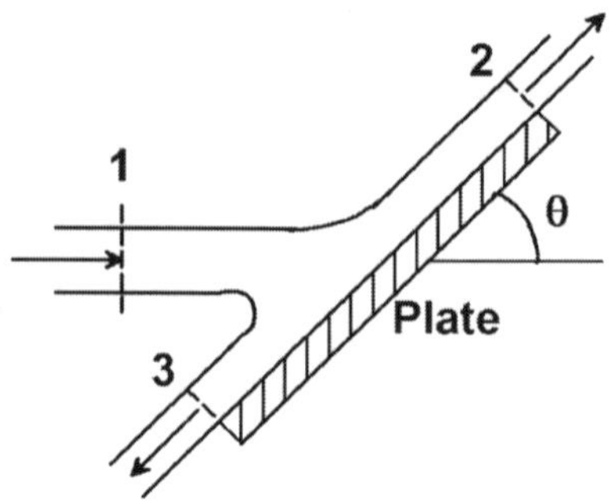

Solution:

Given:

Volumetric flow rae, Q = 0.012 m³/s

Cross- sectional area, $V = \frac{Q}{A} = \frac{0.012}{6\times10^{-4}} = 20\,\text{m/s}.$

Now, velocity of water jet, $V = \frac{Q}{A} = \frac{0.012}{6\times10^{-4}} = 20\,\text{m/s}.$

Since, there is no frictional loss and pressure at all points is equal to atmospheric pressure ~~are equal~~, therefore from Bernoulli's theorem, velocity of the two streams will be same i.e. $V_2 = V_3 = V = 20$ m/s.

Moreover, since there is no friction force along the surface of the plate, therefore net force along the surface of the plate will be zero.

Therefore, from momentum equation, we have, $0 = \rho Q_2 V - \rho Q_3 V - \rho QV\cos\theta$

i.e. $Q_2 - Q_3 = Q\cos\theta$ ... (1)

And, $Q_2 + Q_3 = Q$ ... (2)

Solving (1) and (2) we get, $Q_2 = \frac{Q}{2}(1+\cos\theta)$ and $Q_3 = \frac{Q}{3}(1-\cos\theta)$

$$\text{Therefore, } \frac{(1+\cos\theta)}{(1-\cos\theta)} = 3$$

$$\cos\theta = \frac{1}{2}$$

$$\sin\theta = \frac{\sqrt{3}}{2}$$

Now, force exerted by the water jet in the direction normal to plate is given by,

$$F = \rho QV\sin\theta = 1000\times0.012\times20\times\frac{\sqrt{3}}{2} = 207.8\text{ N}$$

Hence, the answer is 207.8 N.

GATE 2014
FLUID MECHANICS

Q1 – Q9 carry one mark each.

Q1. **A dam with a curved shape is shown in the figure. The cross-sectional area of the dam (shaded portion) is 100 m^2 and its centroid is at x =10 m. The vertical component of the hydrostatic force, F_z, is acting at a distance x_p. The value of x_p is _____ m.**

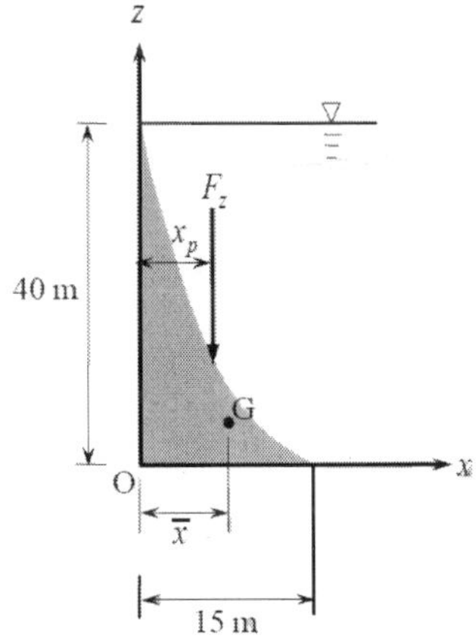

Solution: ______________________________

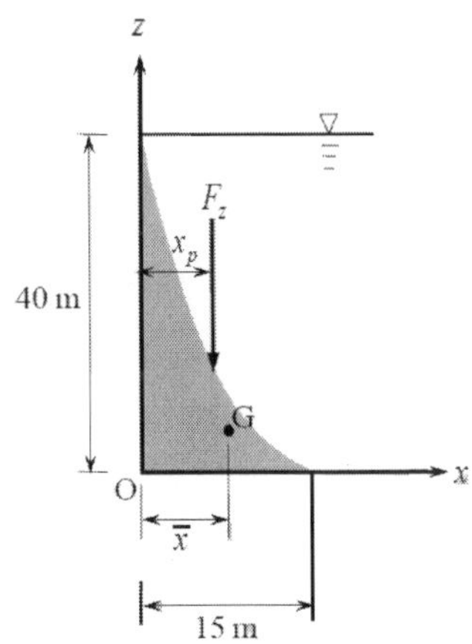

Given:

Cross-sectional area of the dam = 100 m^2

Centroid of the cross-sectional area = 10 m

Area of rectangle ABCO = $\overline{x} = \frac{15}{2}$ m = 7.5 m

Area of shaded portion, A_1 = 100 m^2

Area of portion BCA, A_2 = (600 – 100) m^2 = 500 m^2

Now, centroid of rectangular ABCO,

Now, $\overline{x} = \frac{A_1 \overline{\overline{x_1}} + A_2 \overline{\overline{x_2}}}{A_1 + A_2}$; where is the x-coordinate of the centroid of area ABC

$$\frac{15}{2} = \frac{100 \times 10 + 500 \times \overline{x2}}{100 + 500}$$

$$\overline{x}_2 = 7_m$$

Since vertical force act through the center of the portion ABC, hence

$$x_p = \overline{x}_2 = 7m$$

Hence, the answer is 7 m.

Q2. For an unsteady incompressible fluid flow, the velocity field is $\vec{V} = (3x^2 + 3)t\,\hat{\imath} - 6xyt\,\hat{\jmath}$ where *x*, *y* are in meters and *t* is in seconds. Acceleration in m/s² at the point *x* = 10 m and *y* = 0 and at *t*=0, as measured by a stationary observer is

(A) 303 (B) 162 (C) 42 (D) 13

Solution:

Given:

Velocity field, $\vec{V} = (3x^2 + 3)t\,\hat{\imath} - 6xyt\,\hat{\jmath}$

$\therefore u = (3x^2 + 3)t$ and $v = -6xyt$

Hence the acceleration along the *x*-direction

$$a_x = \frac{\partial u}{\partial t} + u\frac{\partial u}{\partial x} + v\frac{\partial u}{\partial y}$$

$$= (3x^2 + 3) + (3x^2 + 3)t \times 6xt + (-6xyt) \times 0$$

$$= (3x^2 + 3) + 18x^3t^2 + 18xt^2$$

Also, the acceleration along the x-direction

$$a_y = \frac{\partial v}{\partial t} + u\frac{\partial v}{\partial x} + v\frac{\partial v}{\partial y}$$

$$= -6xy + \left(3x^2 + 3\right)t \times 0 + \left(-6xyt\right) \times \left(-6xt\right)$$

$$= -6xy + 36x^2yt^2$$

$\therefore$ At $x = 10m, y = 0$ and $t = 0, a_x = 3\times(10)^2 + 3 + 0 + 0 = 303\,\text{m/s}^2, a_y = 0$

Therefore the resultant acceleration, $a = \sqrt{a_x^2 + a_y^2} = 303\frac{\text{m}}{\text{s}^2}$

Hence, the answer is (A).

Q.3 For an incompressible flow, the existence of components of acceleration for different types of flow is described in the table below.

	Type of Flow		Components of Acceleration
P:	Steady and uniform	1:	Local exists, convective does not exist
Q:	Steady and non-uniform	2:	Both exist
R:	Unsteady and uniform	3:	Both do not exist
S:	Unsteady and non-uniform	4:	Local does not exist, convective exists

Which one of the following options connecting the left column with the right column is correct?

(A) P – 1; Q – 4; R – 3; S – 2 (B) P – 4; Q – 1; R – 2; S – 3
(C) P – 3; Q – 2; R – 1; S – 4 (D) P – 3; Q – 4; R – 1; S – 2

Solution:

P: for steady flow, convective acceleration exists but local acceleration does not exist. For uniform flow, local acceleration exists but convective acceleration does not exist i.e. for steady and uniform flow, neither convective acceleration exists nor local acceleration.

Q: for steady and non-uniform flow, local acceleration does not exist but convective acceleration exists.

R: for unsteady and uniform flow local acceleration exists but convective acceleration does not exist.

S: for unsteady and non-uniform flow both exist.

$\therefore$ P $\rightarrow$ 3

Q $\rightarrow$ 4

R $\rightarrow$ 1

S $\rightarrow$ 2

Hence, the answer is (D).

Q4. Velocity in a two-dimensional flow field is specified as: $u = x^2 y$; $v = -y^2 x$. The magnitude of the rate of angular deformation at a location (x = 2 m and y = 1 m) is ____ s^{-1}.

Solution:

Given:

$u = x^2y$ and $v = -y^2x$

∴ Rate of angular deformation is given by,

$$\dot{\gamma} = \left(\frac{\partial v}{\partial x} + \frac{\partial u}{\partial y}\right) = -y^2 + x^2$$

$$\therefore \dot{\gamma}(\text{at } x = 2 \text{ m} \quad \text{and} \quad y = 1m) = \left[-(1)^2 + (2)^2\right]\text{s}^{-1} = 3\text{s}^{-1}$$

Hence, the answer is 3.

Q5. For a plane irrotational flow, equi-potential lines and streamlines are

(A) parallel to each other.
(B) at an angle of 90 to each other.
(C) at an angle of 45 to each other.
(D) at an angle of 60 to each other.

Solution:

For a plane irrotational flow, equipotential lines and streamlines are at an angle of 90 to each other.

Hence, the answer is (B).

Q6. Flow around a Rankine half-body is represented by the superposition of

(A) source and vortex flows.
(B) source and uniform flows.
(C) vortex and uniform flows.
(D) source, vortex and uniform flows.

Solution:

Flow around a Rankine half-body is represented by the superposition of source and uniform flow.

Hence, the answer is (B).

Q7. It is required to carry out model studies on a boat having a characteristic length of 3.6 m and travelling at a speed of 3 m/s. Assume the acceleration due to gravity as 10 m/s² and neglect the effects due to viscous and surface tension forces. The value of appropriate non-dimensional number is ____.

Solution:

Given:

Characteristic length of the boat, *Lp*= 3.6m

Speed of the boat, *Vp*= 3 m/s

For free surface flow, non-dimensional number is Froude number which is given by,

$$F_r = \frac{V_p}{\sqrt{L_p g}}$$
$$= \frac{3}{\sqrt{3.6 \times 10}}$$
$$= 0.5$$

Hence, the answer is 0.5.

Q8. Which one of the following velocity profiles typically represents a fully developed incompressible, turbulent flow in a pipe?

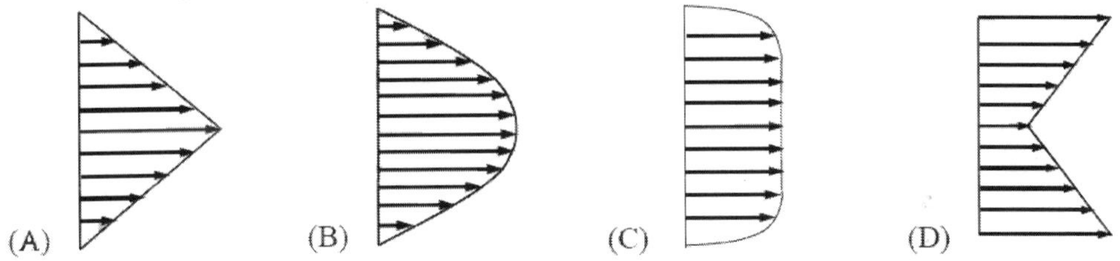

Solution:

A typical velocity profile for a fully developed incompressible, turbulent flow is represented by figure (C).

Q9. Consider an incompressible, laminar flow past a circular cylinder of diameter d. The flow is uniform at the far upstream. Which one of the following figures typically represents the wake velocity profile just downstream of the cylinder?

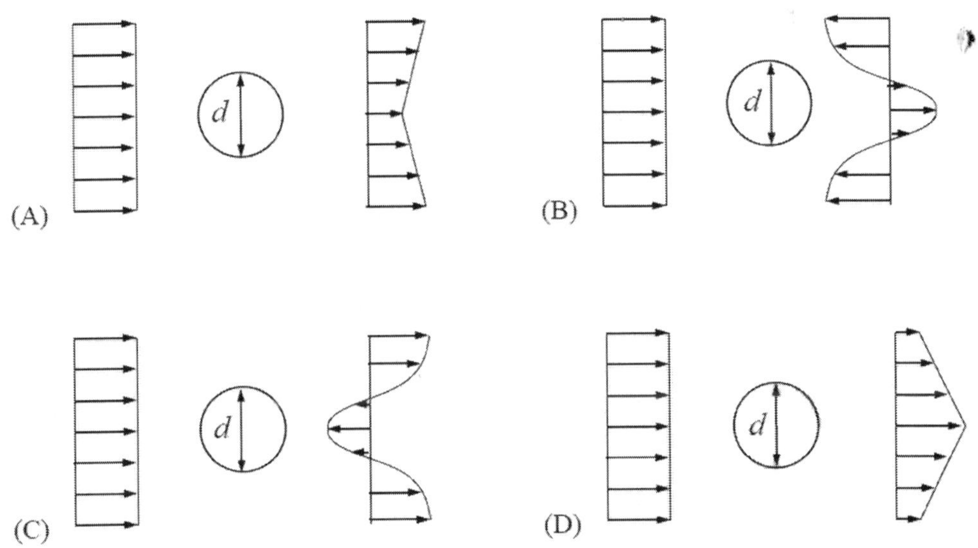

Solution:

For an incompressible laminar flow past a circular cylinder, the wake velocity profile is represented by figure (C).

Q. 10 – Q. 22 carry two marks each.

Q10. A container of square cross-section is partially filled with a liquid of density. The cylinder is intended to float in another liquid of density as shown in the figure. The distance between metacentre and centre of buoyancy is, where I and V_{sub} are area moment of inertia of the cross-section and submerged volume, respectively. Neglect the weight of the container.

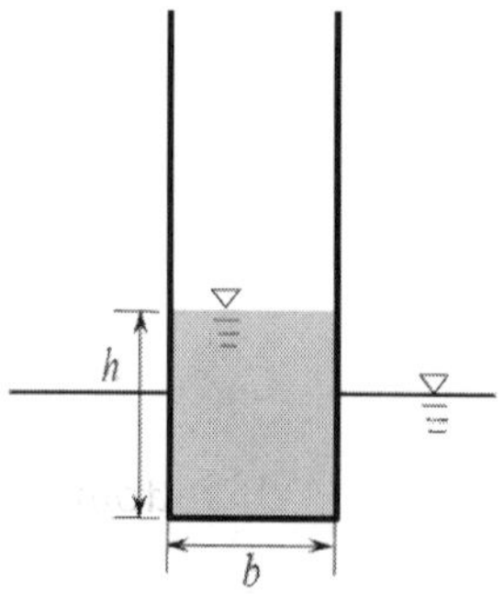

Which one of the following is the correct condition for stability?

(A) $\frac{\rho_2}{6\rho_1}\frac{b}{h}-\frac{h}{b}\left(1-\frac{\rho_1}{\rho_2}\right)>0$ (B) $\frac{\rho_2}{6\rho_1}\frac{b}{h}-\frac{h}{b}\left(1+\frac{\rho_1}{\rho_2}\right)>0$

(C) $\frac{\rho_2}{6\rho_1}\frac{b}{h}+\frac{h}{b}\left(1-\frac{\rho_1}{\rho_2}\right)>0$ (D) $\frac{\rho_2}{6\rho_1}\frac{b}{h}+\frac{h}{b}\left(1+\frac{\rho_1}{\rho_2}\right)>0$

Solution:

Given:

Density of liquid filled in the container, $= \rho_1$

Density of liquid in which container floats $= \rho_2$

Distance between metacenter and center of buoyancy $= \dfrac{I}{\forall_{sub}}$

Now. For the container to be in equilibrium (stable), weight of the liquid filled in the container (neglecting the weight of container) must be equal to the buoyancy force exerted by another liquid.

i.e. $\rho_1 \forall_1 g = \rho_2 \forall_2 g$

where, is the volume of that portion of the container, which is immersed in the liquid.

$\therefore \rho_1 Ah = \rho_2 Ah'$,

where A is the cross-sectional area of the container and is the submerged depth of the container.

Therefore, $h' = \frac{\rho_1}{\rho_2} h$ ...(1)

Moreover, for the container to be in stable equilibrium, metacentric higher must be greater than zero.

i.e. $GM = \frac{I}{\forall_{sub}} - BG > 0$

where, GM is the metacentric height and BG is the distance between center of buoyancy and center of gravity.

Now, distance of center of gravity from bottom = $\frac{h}{2}$

And distance of center of buoyancy from bottom $= \frac{h'}{2} = \frac{\rho_1}{\rho_2}\frac{h}{2}$ [from eq. (1)]

$$\therefore BG = \frac{h}{2} - \frac{h'}{2} = \frac{h}{2} - \frac{\rho_1}{\rho_2}\frac{h}{2}\left(1 - \frac{\rho_1}{\rho_2}\right)\frac{h}{2}$$

Now area moment of inertia of the cross-section, $I = \frac{1}{12} \times b \times b^3$

And submersed volume, $\forall_{sub} = A \times h' = b^2 \times \frac{\rho_1}{\rho_2} h$

$$\therefore \frac{I}{\forall_{sub}} = \frac{\frac{1}{12} \times b \times b^3}{b^2 \times \frac{\rho_1}{\rho_2} h} = \frac{b^2}{12h}\frac{\rho_2}{\rho_1}$$

Hence, GM $= \frac{I}{\forall_{sub}} - BG = \frac{b^2}{12h}\frac{\rho_2}{\rho_1} - \left(1 - \frac{\rho_1}{\rho_2}\right)\frac{h}{2} > 0$

$$\frac{b}{2}\left[\frac{\rho_2 b}{6\rho_1 h} - \left(1 - \frac{\rho_1}{\rho_2}\right)\frac{h}{b}\right] > 0$$

$$\frac{\rho_2}{6\rho_1 h} - \frac{h}{b}\left(1 - \frac{\rho_1}{\rho_2}\right) > 0$$

Hence, the answer is (A).

Q11. In a steady state two-dimensional potential flow field due to a point source, the acceleration of a particle at a distance *r* from the point source is

(A) proportional to r^{-1}
(B) proportional to r
(C) a constant
(D) proportional to r^{-3}

Solution:

Strength of a point source is given by,

$Q = 2r.1.V_r$

where Vr is the radial velocity.

$\therefore V_r = \dfrac{Q}{2\pi r}$ and tangential velocity for source, $V_\theta = 0$

Now the acceleration in radial direction is given by

$$a_r = V_r \frac{\partial V_r}{\partial r} + \frac{V_\theta}{r}\frac{\partial V_r}{d\partial}$$

$$= \frac{Q}{2\pi r}\frac{\partial}{\partial r}\left(\frac{Q}{2\pi r}\right) + 0$$

$$= \frac{Q}{2\pi r} \times \frac{Q}{2\pi}\frac{d}{dr}\left(\frac{1}{r}\right)$$

$$= \frac{Q^2}{4\pi^2 r} \times \left(-\frac{1}{r^2}\right)$$

$$= \frac{-Q^2}{4\pi^2 r^3}$$

$$Kr^{-3}$$

$$\therefore a_r \infty r^{-3}$$

Hence, the answer is (D).

Q12. Velocity in a two-dimensional flow at time *t* and location (*x*, *y*) is described as: $V = 3t^2\hat{i} + (x-1)\hat{j}$. The equation for the path line of a particle passing through the point (1, 0) a *t* =0 is

(A) $x^4 - 4y^3 = 0$
(B) $(x-1)^3 - 2y^4 = 0$
(C) $(x-1)^4 - 64y^3 = 0$
(D) $(x+1)^4 - 16y^3 = 0$

Solution:

$V = 3t^2\hat{i} + (x-1)\hat{j}.$

Hence,

$u = 3t^2$ and $v = x - 1$

$$\frac{dx}{dt} = 3t^2 \qquad \text{...(i)}$$

and $$\frac{dy}{dt} = x - 1 \qquad \text{...(ii)}$$

Integrating eq. (i) gives

$$x = 3 \times \frac{t^3}{3} + C_1 \qquad \text{...(iii)}$$

Now, at $t = 0$, $x = 1$. at $C_1 = 1$, . Hence eq. (iii) gives . Therefore eq. (iii) can be recast as

$$x = t^3 + 1 \qquad \text{...(iv)}$$

Also from eq. (ii) and eq. (iv) we get

$$\frac{dy}{dt} = t^3$$ [Since from $x - 1 = t^3$, eq. (iv)]

$$y = \frac{t^4}{4} + C_2 \qquad \text{...(v)}$$

Since at , hence from eq. (v) we get . Therefore eq. (v) becomes

$$y = \frac{t^4}{4} \qquad \text{...(vi)}$$

Now from eq. (iv) we get

$$t = (x - 1)^{1/3} \qquad \text{...(vii)}$$

Further from eq. (v)

$$t = (4y)^{1/4} \qquad \text{...(viii)}$$

Eliminating from eq. (vii) and eq. (viii) we obtain

$$(x = 1)^{1/3} = (4y)^{1/4}$$

$$x - 1 = (4y)^{\frac{3}{4}}$$

$$(x - 1)^4 - 64y^3 = 0$$

Hence, the answer is (C).

Q13. The gravity driven flow over a hump of height h in a canal is shown in the figure. The height of the free surface from the canal bed at upstream of the hump is H. The free surface height reduces to above the hump.

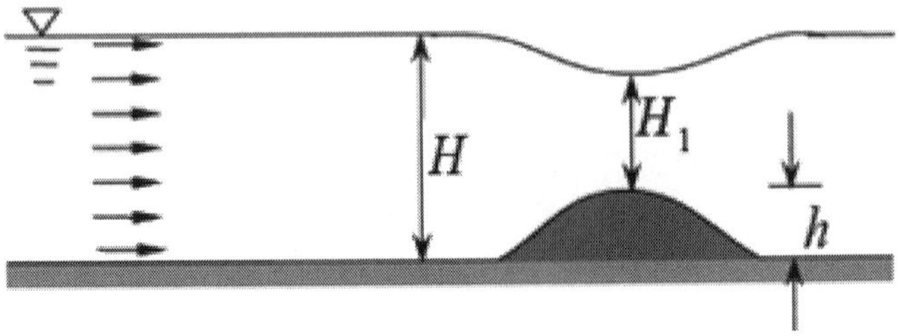

Assuming the canal bed to be horizontal, the discharge per unit width is given by

(A) $\sqrt{\dfrac{2g(h-h_1-h)}{\dfrac{1}{H_1^2}-\dfrac{1}{H^2}}}$ (B) $\sqrt{\dfrac{2gh}{\dfrac{1}{(H_1+h)^2}-\dfrac{1}{H^2}}}$

(C) $\sqrt{\dfrac{2g(H-H_1)}{\dfrac{1}{(H_1+h)^2}-\dfrac{1}{H^2}}}$ (D) $\sqrt{\dfrac{2g(H-H_1)}{\dfrac{1}{H_1^2}-\dfrac{1}{H^2}}}$

Solution:

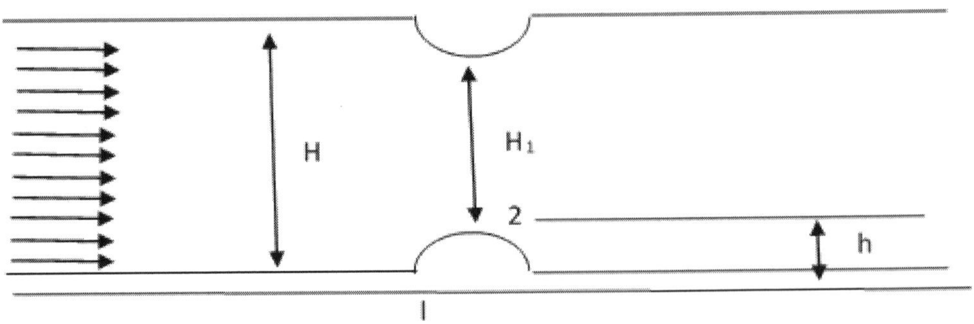

Applying Bernoulli's equation between point 1 and 2, we have

$$\frac{p_1}{\rho g}+\frac{V_1^2}{2g}+Z_1=\frac{P_2}{\rho g}+\frac{V_2^2}{2g}+Z_2$$

$$\frac{\rho g H}{\rho g}+\frac{V_1^2}{2g}=\frac{\rho g H_1}{\rho g}+\frac{V_2^2}{2g}+h$$

$$H+\frac{V_1^2}{2g}=H_1+\frac{V_2^2}{2g}+h \quad \text{... (1)}$$

Now from the continuity equation we have

$A_1V_1 = A_2V_2$

or, $H \times 1 \times V_1 = H_1 \times 1 \times V_2$ [since the width of the canal is unity]

$$V_1 = \frac{H_1}{H} V_2 \qquad \ldots(2)$$

Using eq. (1) and eq. (2) we get

$$H + \frac{\left(\frac{H_1}{H} V_2\right)^2}{2g} = H_1 + \frac{V_2^2}{2g} + h$$

$$H + \left(\frac{H_1}{H}\right)^2 \frac{V_2^2}{2g} = H_1 + \frac{V_2^2}{2g} + h$$

$$H + \left(\frac{H_1}{H}\right)^2 \frac{V_2^2}{2g} = \frac{V_2^2}{2g} + (h + H_1)$$

$$\frac{V_2^2}{2g}\left(\frac{H_1^2}{H^2} - 1\right) = (h + H_1) - H$$

$$V_2^2 = \frac{\{(h + H_1) - H\}}{\left(\frac{H_1^2}{H^2} - 1\right)}$$

$$V_2^2 H_1^2 = \frac{2g\{(h + H_1 + H)\}}{\left(\frac{H_1^2 - H^2}{H^2}\right)} H_1^2$$

$$Q^2 = \frac{2g\{(h + H_1 - H)\}}{\left(\frac{H_1^2 - H^2}{H^2 H_1^2}\right)}$$

$$Q = \sqrt{\frac{2g\{H - H_1 - h\}}{\frac{1}{H_1^2} - \frac{1}{H^2}}}$$

Hence, the answer is (A).

Q14. Steady state incompressible flow through a pipe network is shown in the figure. Inlets marked as (1), (2) and (3) and exit marked as (4), are shown with their respective diameters. The exit flow rate at (4) is 0.1 m³/s. A 20% increase in flow rate through (3) results in a 10% increase in flow rate through (4). The original velocity through inlet (3) is _____ m/s.

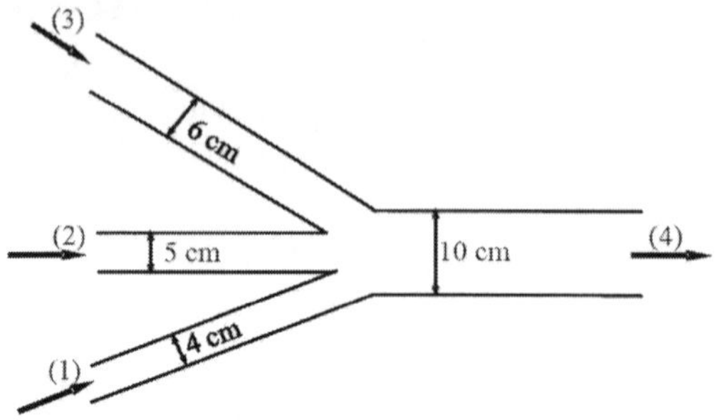

Solution:

Let,

Flow rate through section 1 = Q_1

Flow rate through section 2 = Q_2

Flow rate through section 3 = Q_3

Flow rate through section 4 = Q_4

Now, we have, $Q_1 + Q_2 + Q_3 = Q_4$ and $Q_4 = 0.1$ m³/s

Now, A/q $Q'_3 = Q_3 + 0.20\ Q_3 = 1.20\ Q_3$

$Q'_4 = Q_4 + 0.10\ Q_4 = 1.10\ Q_4$

$\therefore Q_1 + Q_2 + Q'_3 = Q'_4 = 1.10\ Q4$ → (i)

And $Q_1 + Q_2 + Q_3 = Q_4$ → (ii)

Subtracting (ii) from (i),

$Q'_3 - Q_3 = 1.10\ Q_4 - Q_4$

$1.20\ Q_3 - Q_3 = 0.10\ Q_4$

$0.20\ Q_3 = 0.10\ Q_4$

$Q_3 = 0.5\ Q_4$

$Q_3 = 0.5 \times 0.1$ m³/s = 0.05 m³/s

Now, we have $Q_3 = A_3 V_3 = \dfrac{\pi d_3^2}{4} \times V_3$

Hence, $V_3 = \dfrac{4Q_3}{\pi d_3^2} = \dfrac{4 \times 0.05}{3.14 \times 0.06 \times 0.06} = 17.69\,\text{m/s}$

Hence, the answer is 17.69 m/s.

Q15. A reducing elbow is used to deflect water upward by 30 as shown in the figure. The mass flow rate at the inlet is 14 kg/s. Water is entering at a gauge pressure of 200 kPa and exits to the atmosphere. The cross-sectional area is 113 cm² at the inlet and 7 cm² at the exit. Density of water and acceleration due to gravity are 1000 kg/m³ and 10 m/s², respectively. Magnitude of *x* – component of the water force on the elbow is ____ N.

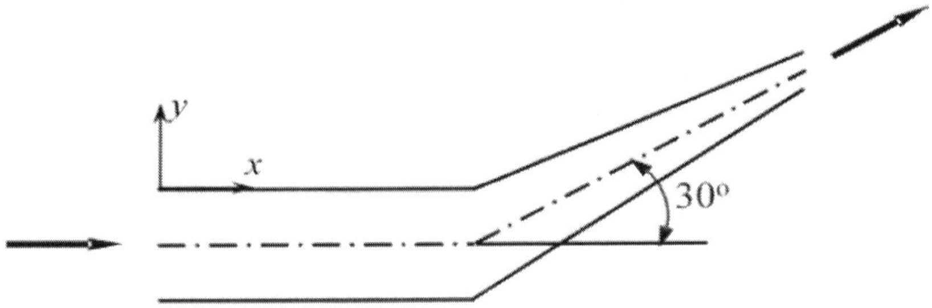

Solution:

Given:

Mass flow rate of the inlet, = 14 kg/s

Gauss pressure at inlet, $p_1 = 200$ kPa = 2×10^5 Pa

Pressure at exit, $p_2 = 0$ (gauge)

Cross-sectional area at inlet, $A_1 = 113$ cm² = 113×10^{-4} m²

Cross-sectional area at outlet, $A_2 = 7$ cm² = 7×10^{-4} m²

Density of water, ρ = 1000 kg/m³

$$\text{Now, } \dot{m} = \rho A_1 V_1 = \rho A_2 V_2$$

$$\text{Hence } V_1 = \frac{\dot{m}}{\rho A_1} = \frac{14}{1000\times113\times10^{-4}}\,\text{m/s} = 1.24\,\text{m/s}$$

$$\text{and } V_2 = \frac{\dot{m}}{\rho A_2} = \frac{14}{1000\times7\times10^{-4}}\,\text{m/s} = 20\,\text{m/s}$$

Let F_x be the force exerted by the bend on the jet in the *x*-direction. Then the momentum equation along the *x*-direction can be written as

$$\dot{m}\{V_2\cos(30°) - V_1\} = p_1A_1 - p_2A_2\cos(30°) + F_x$$

$$F_x = \dot{m}\{V_2\cos(30°) - V_1\} + p_2A_2\cos(30°) - p_1A_1$$

$$= 14\times\{20\times\cos(30°) - 1.24\} - 200\times10^3\times113\times10^{-4}\,\text{N}$$

$$= -2037.87\ \text{N}$$

Hence, the answer is 2034.87 N .

Q16. A source with a strength of k_1 and a vortex with a strength of k_2 are located at the origin. The resultant velocity at a radial distance r from the origin due to the superposition of the source and vortex is expressed as

(A) $\frac{k_1 + k_2}{r}$ (B) $\frac{\sqrt{k_1^2 + k_2^2}}{r}$ (C) $\frac{\sqrt{k_1^2 - k_2^2}}{r}$ (D) $\frac{k_1 - k_2}{r}$

Solution:

Given:

Strength of the source = k_1 and strength of the vortex = k_2

Now, for source, we have $Q = AV_r$

$Q = 2\pi r X 1 X V_r$

$V_r = Q/2\pi r$

where, Q = volumetric flow rate in the radial direction through unit depth

and V_r = radial velocity

Now, $Q/2\pi$ is called the strength of the source.

∴ We have $Q/2\pi = k_1$

∴ $V_r = k_1/r$ ----- (i)

Now, for vortex flow we have, $V_\theta = \Gamma/2\pi r$, where Γ is circulation of the flow.

∴ $V_\theta = k_2/r$, where $\Gamma/2\pi$ is strength of vortex.

∴ Superposition of source and vertex is given by,

$$V = \sqrt{\left(\frac{k_1}{r}\right)^2 + \left(\frac{k_2}{r}\right)^2}$$

$$V = \frac{\sqrt{k_1^2 + k_2^2}}{r}$$

Hence, the answer is (B).

Q17. Velocity potential for an incompressible fluid flow is given as: $\varnothing = 2(x^2 + 2y - y^2)$. Assume the value of stream function at the origin to be zero. The value of stream function at [(x, y)≡(2,2)] is ____.

Solution:

Velocity potential for the incompressible fluid flow is given by,

$\varnothing = 2(x^2+2y-y^2)$

∴ We have, $u = \frac{\partial \varnothing}{\partial x} = 4x$

And $v = \frac{\partial \varnothing}{\partial y} = 4 - 4y$

Now, $u = \frac{\partial \psi}{\partial y}$ and $v = -\frac{\partial \psi}{\partial x}$

$\therefore \frac{\partial \psi}{\partial y} = 4y$ ...(1)

and $-\frac{\partial \psi}{\partial x} = 4 - 4y$ ...(2)

Hence from eq. (1), we get $\psi = 4xy + f(x)$ ...(3)

And from eq. (2) $\psi = 4xy - 4x + C$ ...(4)

Hence, combining eq. (3) and eq. (4) we can write

$\psi = 4xy - 4x + C$ ...(5)

Now at $(x,y) = (0,0)$. Hence from eq. (5), and the equation for the streamline is

$\psi = 4xy - 4x$

$\therefore \emptyset$ at $[(x,y) = (2,2)] = 4 \times 2 \times 2 - 4 \times 2 = 16 - 8 = 8$

Hence, the answer is 8.

Q18. The model of a conduit is scaled to 1/100 of the actual size. Seawater is used in the prototype and fresh water is used in the model. Velocity in the prototype is 0.5 m/s. Density and dynamic viscosity of the seawater are 1025 kg/m^3 and 1.07 × 10^{-3} kg/m-s, respectively. Density and dynamic viscosity of fresh water are 1000 kg/m^3 and 1 × 10^{-3} kg/m-s, respectively. Assume the viscous forces to be dominant. The velocity to be maintained in the model to ensure dynamic similarity is____ m/s.

Solution:

Given: $\frac{L_m}{L_p} = \frac{1}{100}$

where, L_m = length of the model

L_p = length of the prototype

Velocity of the prototype $Vp = 0.5$ m/s

Density of the prototype $\rho_p = 1025$ kg/m^3

Dynamic viscosity of the prototype $\mu_p = 1.07 \times 10^{-3}$ kg/m-s

Viscosity of the model $\mu_m = 1 \times 10^{-3}$ kg/m-s

Density of the model $\rho_m = 1000$ kg/m^3

Now if viscous force is predominant, for dynamic similarity between model and prototype we have

$Re_m = Re_p$

i.e. $\frac{\rho_m V_m L_m}{\mu_m} = \frac{\rho_p V_p L_p}{\mu_p}$

$$V_m = \left(\frac{\rho_p}{\rho_m}\right) \times \left(\frac{L_p}{L_m}\right) \times \left(\frac{\mu_m}{\mu_p}\right) \times V_p$$

$$V_m = \left(\frac{1025}{1000}\right) \times (100) \times \left(\frac{1 \times 10^{-3}}{1.07 \times 10^{-3}}\right) \times 0.5\,\text{m/s} = 47.89\,\text{m/s}$$

Hence, the answer is 47.89 m/s.

Q19. A fluid is flowing through a pipe of circular cross-section. Reynolds number of the flow is 1600. The head loss over a 45 m length of the pipe is 0.6 m. The average flow velocity of the fluid is 1 m/s and the acceleration due to gravity is 10 m/. The diameter of the pipe is ____ m.

Solution:

Given:

Reynolds number Re = 1600

Heat loss $h_L = 0.6$ m

Length of the pipe $L = 45$ m

Average velocity, $V = 1$ m/s

Now the heat loss through pipe of diameter, d over a length L is given by

$$h_L = f \frac{L}{d} \frac{V^2}{2g}$$

where f is Darcy's friction factor which for a laminar flow is given by

$f = 64/\text{Re}$

$f = 64/1600 = 0.04$

$$h_L = f \frac{L}{d} \frac{V^2}{2g}$$

$0.6 = 0.04$

$d = 0.15$ m

Hence, the answer is 0.15 m.

Q20. Consider a laminar flow over a flat plate of width w. At Section 1-1, the velocity profile is uniform as shown in the figure. The x-direction velocity profile at Section 2-2 is given by $\frac{u}{U} = 2\frac{y}{\delta} = \frac{y^2}{\delta^2}$ where δ is the boundary layer thickness.

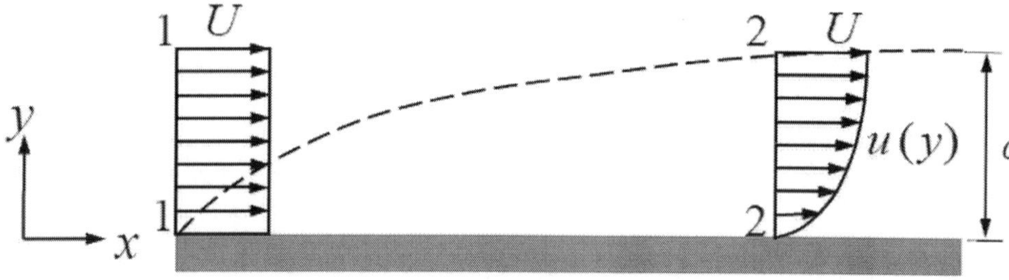

The volume flow rate through Section 2-2 is given by

(A) $\frac{1}{2}Uw\delta$ (B) $\frac{1}{3}Uw\delta$ (C) $Uw\delta$ (D) $\frac{2}{3}Uw\delta$

Solution:

Given:

The velocity profile is given by,

$u/U = 2\,(y/\delta) - (y/\delta)^2$

$u = [2\,(y/\delta) - (y/\delta)^2]\,U$

Now, the volumetric flow rate is given by

$Q = u \times A$

where, A is the cross-sectional area. Hence the flow rate through an infinitesimally small area $dA=wdy$ is given by

$dQ = u \times wdy$

where, dQ = volumetric through a small ship of thickness dy and width w.

$\therefore dQ = [2\,(y/\delta) - (y/\delta)^2]\,Uwdy$

$$Q = 2\frac{Uw}{\delta}\int_0^\delta ydy = \frac{uw}{\delta}\int_0^\delta y^2 dy$$

$$= 2\frac{Uw}{\delta}\left[\frac{y^2}{2}\right]_0^\delta - \frac{uw}{\delta^2}\left[\frac{y^3}{3}\right]_0^\delta$$

$$uw\delta - \frac{uw\delta}{3\delta}$$

$$Q = \frac{2}{3}uw\delta$$

Hence, the answer is (D).

Q21. A cube of weight W and side a falls at a constant speed in a medium as shown in the figure. If the medium is air (mass density = ρ_{air}) let U_{air} be the velocity of the cube. If the medium is water (mass density = ρ_{water}, let U_{water} be the $\frac{U_{air}}{U_{water}}$ velocity of the cube.

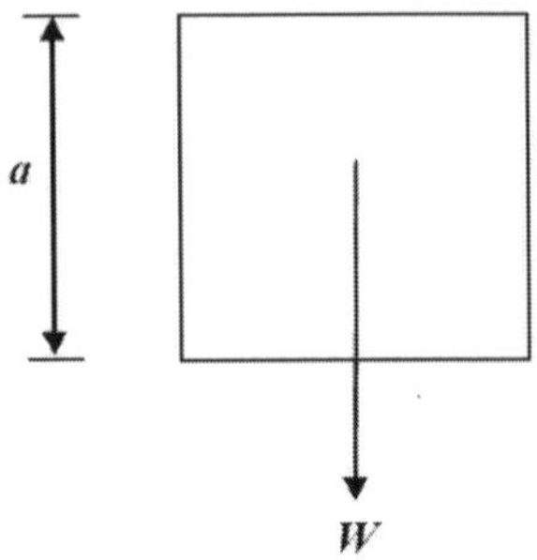

Neglecting the buoyancy force and assuming drag coefficient to be same for both cases, the ratio of velocities, is given by

(A) $\frac{\rho_{air}}{\rho_{water}}$ (B) $\sqrt{\frac{\rho_{air}}{\rho_{water}}}$ (C) $\sqrt{\frac{\rho_{water}}{\rho_{air}}}$ (D) 1

Solution:

Given:

Weight of the cube = W

Mass density of air = ρ_{air}

Velocity of cube in air = U_{air}

Mass density of water = ρ_{water}

Velocity of cube in water = U_{water}

Now, neglecting the buoyancy force, for the cube moving at constant velocity,

We have, $W = F_D$

$W = ½\, C_D\, \rho_{\text{air}}\, AU^2_{\text{air}}$ ------- (i)

And $W = ½\, C_D\, \rho_{\text{water}}\, AU^2_{\text{water}}$ ---- (ii)

Dividing (i) by (ii), we have,

$$\frac{U^2_{air}}{U^2_{water}} = \frac{\rho_{water}}{\rho_{air}}$$

$$\left(\frac{U_{air}}{U_{water}}\right) = \sqrt{\frac{\rho_{water}}{\rho_{air}}}$$

Hence, the answer is (C).

Q22. Water is flowing through a venturimeter having a diameter of 0.25 m at the entrance (Station 1) and 0.125 m at the throat (Station 2) as shown in the figure. A mercury manometer measures the piezometric head difference between Stations 1 and 2 as 1.3505 m. The loss of head between these two stations, is 1/7 times the velocity head at the Station 2. Assume the acceleration due to gravity to be 10 m/s². The velocity of water at the throat is ____ m/s.

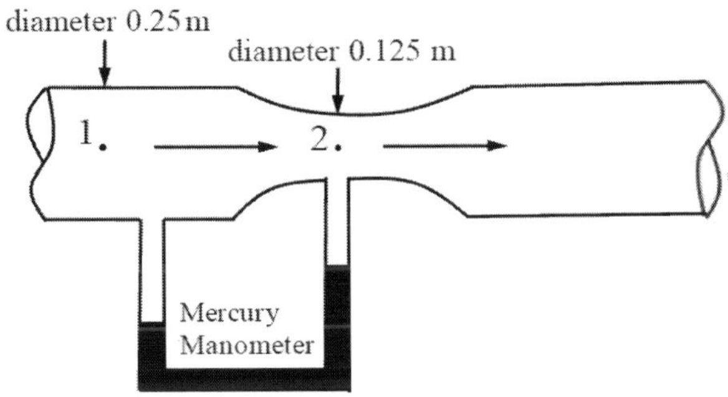

Solution:

Given:

Diameter of pipe, $d_1 = 0.25$ m

Diameter of throat, $d_2 = 0.125$ m

Piezometric head difference $\left(\frac{p_1}{\rho g} + Z_1\right) - \left(\frac{p_2}{\rho g} + Z_2\right) = 1.3505$ m

Heat loss, $h_L = \frac{1}{7}\frac{V_2^2}{2g}$

Now applying Bernoulli's theorem between point 1 and point 2 we have,

$$\frac{p_1}{\rho g} + \frac{V_1^2}{2g} + Z_1 = \frac{p_2}{\rho g} + \frac{V_2^2}{2g} + Z_2 + h_L$$

$$\left(\frac{p_1}{\rho g} + Z_1\right) - \left(\frac{p_2}{\rho g} + Z_2\right) + \frac{V_2^2}{2g} - \frac{V_2^2}{2g} = \frac{1}{7}\frac{V_2^2}{2g}$$

$$1.3505 + \frac{V_1^2}{2g} = \frac{8}{7}\frac{V_2^2}{2g}$$

$$1.3505 = \frac{8}{7}\frac{V_2^2}{2g} - \frac{V_1^2}{2g} \qquad \text{... (1)}$$

Now from continuity equation we have

$A_1V_1 = A_2V_2$

$$\frac{\pi d_1^2}{4}V_1 = \frac{\pi d_2^2}{4}V_2$$
$$d_1^2 V_1 = d_2^2 V_2$$
$$V_1 = \frac{d_2^2}{d_1^2}V_2 = \left(\frac{0.125}{0.25}\right)^2 V_2 = 0.25\,V_2$$

Hence from eq. (1), we have, $1.3505 = \frac{8}{7}\frac{V_2^2}{2g} - \frac{(0.25)^2 V_2^2}{2g}$

$$1.3505 = \left(\frac{8}{7} - 0.0625\right)\frac{V_2^2}{2g}$$
$$\frac{V_2^2}{2\times 10} = \frac{1.3505}{1.08}$$
$$V_2 = 5\,\text{m/s}$$

Hence, the answer is 5 m/s.

GATE 2013
FLUID MECHANICS

Q1. The gauge pressure inside a soap bubble of radius R, with σ denoting the surface tension between the soap solution and air, is.

(A) $\frac{\sigma}{2\pi R}$ (B) $\frac{4\sigma}{R}$ (C) $\frac{2\sigma}{R}$ (D) $\frac{\sigma}{2\pi R}$

Solution:

Consider a soap bubble which is cut into two halves as shown in fig.

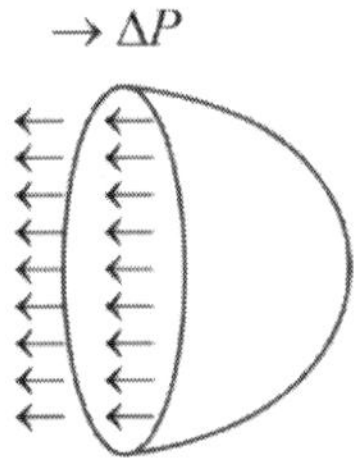

For the soap bubble to be in equilibrium (stable), force by excess pressure inside the soap bubble must counter balance the force due to surface tension.

i.e. $\Delta p \times \pi R^2 = 2 \times 2\pi R$

$$\Delta p = \frac{4\sigma}{R}$$

Hence, the answer is (B).

Q.2 Let M, B and G represent respectively the metacentre, centre of buoyancy and the centre of mass of a floating buoy. Which of the following statements is correct?

(A) M is above G; Buoy unstable (B) B is above G; Buoy stable

(C) M is above G; Buoy stable (D) B is above G; Buoy unstable

Solution:

For a floating body to be in stable equilibrium, metacenter must be above the centre of gravity.

Hence, the answer is (C).

Q3. A reservoir connected to a pipe line is being filled with water, as shown in the Figure. At any time t, the free surface level in the reservoir is h. Find the time in seconds for the reservoir to get filled up to a height of 1 m, if the initial level is 0.2 m ________

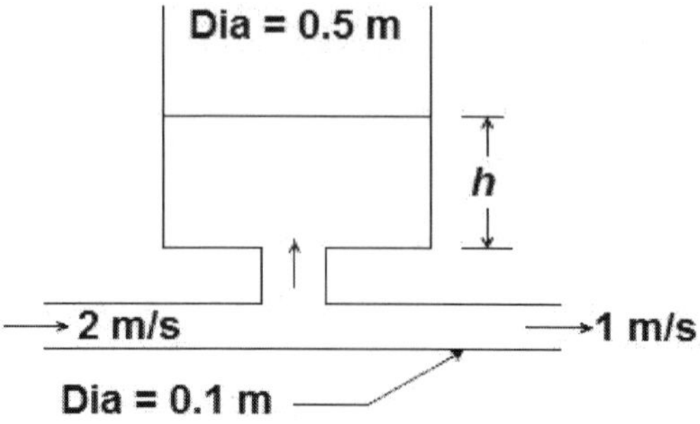

Solution:

If any time t, the free surface level in the reservoir is h, and dh is the rise in water level in same time interval dt, then volumetric flow ratio in the tank is given by

$Q = \frac{Adh}{dt}$, where A is the cross sectional area of the tank.

Now, from continuing equation, we have,

$$Q_1 - Q_2 = Q$$

where Q_1 and Q_2 are the volumetric flow ratio of inner and outer pipe respectively.

$$\therefore A_1V_1 - A_2V_2 = \frac{Adh}{dt}$$

Or $\frac{\pi d_1^2}{4} V_1 - \frac{\pi d_2^2}{4} V_2 = \frac{\pi d^2}{4} \frac{dh}{dt}$

where d_1, d_2 and d are the diameter of inter; outer pipe and tank respectively.
Now, $d_1 = d_2 = 0.1$ m, $d = 0.5$ m and $V_1 = 2$ m/s, $V_2 = 1$ m/s

$$\therefore d_1^2 V_1 - d_2^2 V_2 = d^2 \frac{dh}{dt}$$

$$(0.1)^2 \times 2 - (0.1)^2 \times 1 = (0.5)^2 \frac{dh}{dt}$$

$$\frac{dh}{dt} = 0.04$$

$$dt = \frac{1}{0.04} dh$$

$$\int_0^t dt = 25 \int_{0.2}^1 dh$$

$$t = 25 \times (1 - 0.2) \sec$$

t = 20 sec.

Hence, the answer is 20 sec.

Q.4 Bernoulli's equation is valid for the following type of flow:

(A) Compressible, steady, inviscid (B) Incompressible, steady, viscous

(C) Compressible, unsteady, viscous (D) Incompressible, steady, inviscid

Solution:

Bernoulli's equation is valid for

Incompressible flow

Steady flow

Irrotational flow

Inviscid flow

Hence, the answer is (D).

Q5. If A is the area of a circle of radius r enclosing a plane forced vortex flow, with origin at the centre of the vortex and if ω is the angular velocity, ζ is the vorticity , V is the velocity vector, then the circulation around the contour of the circle is given by

(A) $2\omega A$ (B) $2\zeta A$ (C) $2VA$ (D) 0

Solution:

Vorticity is given by, $\zeta = \frac{circulation(\Gamma)}{Area(A)}$

$\therefore \Gamma = \zeta A$

And vorticity, $\zeta = 2 \times \omega$

Hence, $\Gamma = 2\omega A$

Hence, the answer is (A).

Q6. Flow past a circular cylinder can be produced by superposition of the following elementary potential flows:

(A) Uniform flow, doublet (B) Uniform flow, vortex

(C) Source, vortex (D) Sink, vortex

Solution:

Flow past a circular cylinder can be produced by superposition of uniform flow and doublet.

Hence, the answer is (A).

Q7. Let δ, δ_1 and δ_2 denote respectively the boundary-layer thickness, displacement thickness and the momentum thickness for laminar boundary layer flow of an incompressible fluid over a flat plate. The correct relation among these quantities is

(A) $\delta<\delta_1<\delta_2$ (B) $\delta>\delta_1>\delta_2$ (C) $\delta>\delta_1<\delta_2$ (D) $\delta<\delta_1>\delta_2$

Solution:

Boundary layer thickness is given by, $\frac{\delta}{x}=\frac{5}{\sqrt{\mathrm{Re}\, x}}$

$$\delta=\frac{5x}{\sqrt{\mathrm{Re}\, x}}$$

Displacement thickness is given by

And momentum thickness is given by

Therefore we have, $\delta>\delta_1>\delta_2$

Hence, the answer is (B).

Q8. In the hydrodynamic entry region of a circular duct, the pressure forces balance the sum of

(A) viscous and buoyancy forces (B) inertia and buoyancy forces
(C) inertia and surface tension forces (D) inertia and viscous forces

Solution:

In the hydrodynamic entry region of a circular duct, the pressure forces balance the sum of the inertia and viscous forces.

Hence, the answer is (D).

Q.9 Bodies with various cross-sectional shapes subjected to cross-flow of air are shown in the following figures. The characteristic dimension of all the shapes is the same. The cross-sectional shape with the largest coefficient of drag (i.e. sum of the pressure and skin-friction drags), at any moderately large Reynolds number, is

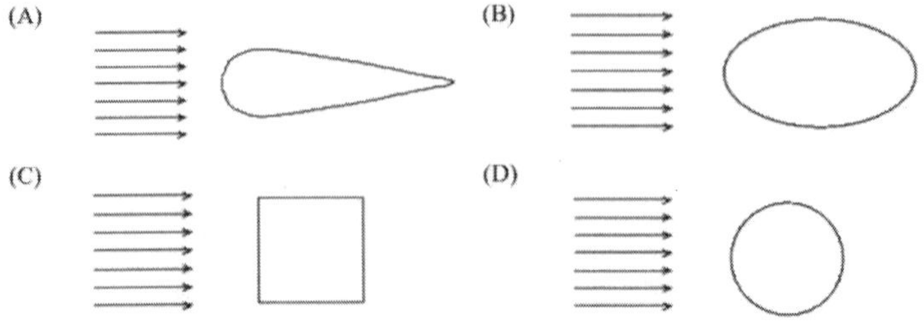

Solution:

At moderating large number, coefficient of drag is independent of Reynolds number, but it depends on the characteristic dimensions and the cross sectional shape. It is observed that for bluff bodies, drag coefficient is higher than that of streamline bodies. Therefore, square being bluff body will have largest drag coefficient.

Hence, the answer is (C).

Q. 10 - Q. 22 carry two marks each.

Q.10 A U-tube of a very small bore, with its limbs in a vertical plane and filled with a liquid of density ρ, up to a height of h, is rotated about a vertical axis, with an angular velocity of ω, as shown in the Figure. The radius of each limb from the axis of rotation is R. Let p_a be the atmospheric pressure and g, the gravitational acceleration. The angular velocity at which the pressure at the point O becomes half of the atmospheric pressure is given by

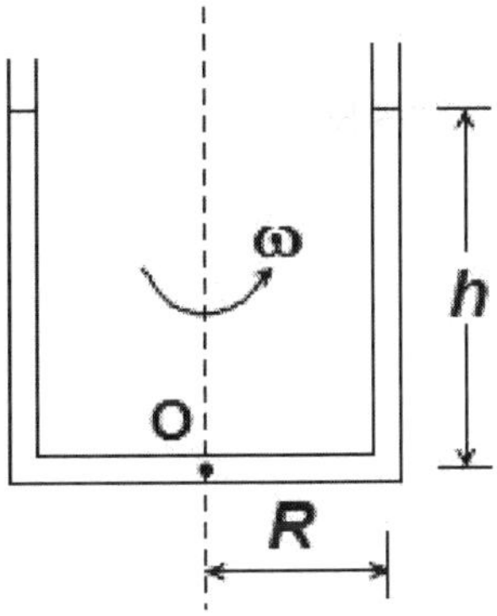

(A) $\dfrac{\sigma}{2\pi R}$ (B) $\sqrt{\dfrac{2(p_a+\rho gh)}{\rho R^2}}$

(C) $\sqrt{\dfrac{p_a+2\rho gh}{2\rho R^2}}$ (C) $\sqrt{\dfrac{p_a+\rho gh}{2\rho R^2}}$

Solution:

If p_1 and p_2 are the pressure of radial distances r_1 and r_2 respectively and ω is the angular velocity, then pressure distribution is given by

$$p_1 - p_2 = \frac{1}{2}\rho\omega^2\left(r_1^2 - r_2^2\right)$$

Now, at $r = 0$, $p_1 = p_a/2$ and at $r = R$, $p_2 = p_a + \rho gh$

$$\frac{p}{2} - p_a - \rho gh = \frac{1}{2}\rho\omega^2\left(0 - R^2\right)$$

$$-\frac{p_a}{2}-\rho gh=-\frac{1}{2}\rho\omega^2R^2$$

$$\frac{1}{2}\rho\omega^2R^2=\frac{p_a+2\rho gh}{2}$$

$$\omega^2=\frac{p_a+2\rho gh}{\rho R^2}$$

$$\omega=\sqrt{\frac{p_a+2\rho gh}{\rho R^2}}$$

Hence, the answer is (A).

Q11. An incompressible fluid at a pressure of 150 kPa (absolute) flows steadily through a two dimensional channel with a velocity of 5 m/s as shown in the Figure. The channel has a 90° bend. The fluid leaves the channel with a pressure of 100 kPa (absolute) and linearly varying velocity profile. V_{max} is four times V_{min}. The density of the fluid is 914.3 kg/m³. The velocity V_{min}, in m/s, is

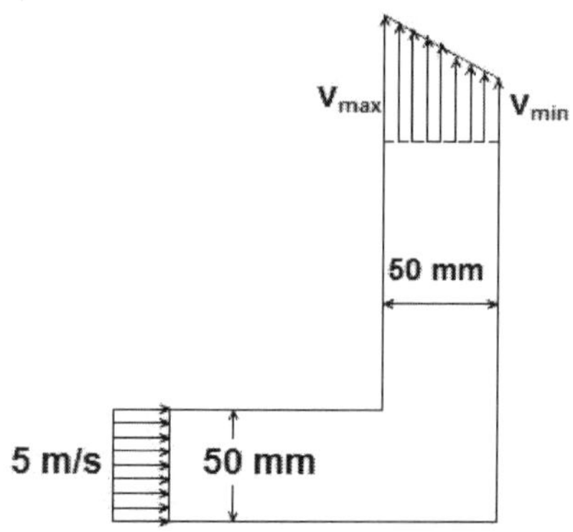

(A) 25 (B) 2.5 (C) 2.0 (D) 0.2

Solution:

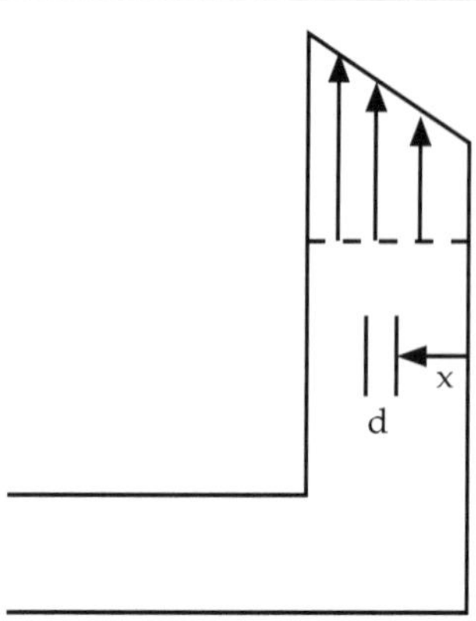

Given:

The inlet velocity of fluid, V=5 ms^{-1}

$V_{max} = 4V_{min}$

Let the velocity profile of the outlet of the channel is given by,

$U= ax + b$ -------- (1)

where a and b are constants and U is the velocity of horizontal distance x. Now from boundary conditions we have,

At $x = 0$: $U = V_{min}$

$V_{min} = a0 + b$

$b = V_{min}$

And at $x = 50$ mm = 0.05 m, $U = V_{max}$

$V_{max} = a0.05+b$

$V_{max} = 0.05a + V_{min}$

$V_{max} - V_{min} = 0.05a$

$4V_{min} - V_{min} = 0.05a$

$3V_{min} = 0.05a$

$a = 60V_{min}$

Therefore, the velocity profile is given by

$U = 60 \text{ X } V_{min} \text{ X } x + V_{min}$ -------- (2)

Now, volumetric flow rate through small cross section with thickness dx and depth unity is given by

$$dQ = Udx.1$$

$$dQ = (60 \times V_{min} \times x + V_{min})dx$$

$$Q = \int_{0.}^{0.05} 60xV_{min}dx + \int_{0}^{0.05} V_{min}dx$$

$$= 60V_{min}\int_{0}^{.05} xdx + V_{min}\int_{0}^{0.05} dx$$

$$= 60V_{min}\left[\frac{x^2}{2}\right]^{0.05} + V_{min}\left[x\right]_{0}^{0.05}$$

$$= 30V_{min}\left[(0.05)^2 - 0\right] + V_{min} \times 0.05$$

$$= 0.075V_{min} + 0.05V_{min}$$

$$Q = 0.0125V_{min}$$

We also have, $Q = A\ V = 0.05\ 1\ 5 = 0.25\ m^3s^{-1}$

Hence, $0.125\ V_{min} = 0.25$

$V_{min} = 2\ ms^{-1}$

Hence, the answer is (C).

Q12. The velocity vector corresponding to a flow field is given, with usual notation, by $\vec{V} = 3x\hat{\imath} + 4xy\hat{j}$. The magnitude of rotation at the point (2,2) in rad/s is

(A) 0.75 (B) 1.33 (C) 2 (D) 4

Solution:

Given: $\vec{V} = 3x\hat{\imath} + 4xy\hat{j}$.

$\therefore u = 3x$ and $v = 4xy$

Now rotation is given by

$$\omega = \frac{1}{2}\left(\frac{\partial v}{\partial x} - \frac{\partial u}{\partial y}\right)$$

$$= \frac{1}{2}(4y - 0) = 2y$$

Hence, at (2,2) = 22 = 4 rad/s

Hence, the answer is (D).

Q13. The stream function for a potential flow field is given by $\psi = x^2 - y^2$. The corresponding potential function, assuming zero potential at the origin, is

(A) $x^2 + y^2$ (B) $2xy$ (C) $x^2 - y^2$ (D) $x - y$

Solution:

The stream function for a given flow field is given by

$\psi = x^2 - y^2$

Now, $u = -\frac{\partial \psi}{\partial y}$ and $v = \frac{\partial \psi}{\partial x}$

$u = 2y$ and $v = 2x$

Also, we have $u = \frac{\partial \varphi}{\partial x}$ and $v = \frac{\partial \varphi}{\partial y}$

$\therefore 2y = \frac{\partial \varphi}{\partial x}$ and $2x = \frac{\partial \varphi}{\partial y}$

$d\varphi = 2ydx$ and $d\varphi = 2xdy$

$\therefore \varphi = 2xy + c$

At $x = 0$ and $y = 0$, = 0

$\therefore 0 = 2\times0\times0 + c$

$c = 0$

$\therefore \varphi = 2xy$

Hence, the answer is (B).

Q14. Fully developed flow of an oil takes place in a pipe of inner diameter 50 mm. The pressure drop per metre length of the pipe is 2 kPa. Determine the shear stress, in Pa, at the pipe wall. ________

Solution:

Given:

Inner diameter, d = 50 mm = 0.05 m

Pressure drop per unit length, $\frac{\Delta p}{L}$ = 2 KPa = 210³ Pa

Now, shear stress at racial distance r from the centerline of pipes is given by,

$$\tau = \left(\frac{\Delta P}{L}\right)\frac{r}{2}$$

∴ Shear stress at wall is given by

$$\tau = \left(\frac{\Delta p}{L}\right)\frac{r}{2} = \left(\frac{\Delta p}{L}\right)\frac{d}{2\times 2} = \left(\frac{\Delta p}{L}\right)\frac{d}{4}$$

$$\therefore \tau = 2\times 10^3 \times \frac{0.05}{4} = 25 \text{ pa.}$$

= 2 10³ = 25 Pa.

Hence, the answer is 25 Pa.

Q15. The Darcy friction factor f for a smooth pipe is given by f = 64/Re for laminar flow and by f = 0.3/Re$^{0.25}$ for turbulent flow, where Re is the Reynolds number based on the diameter. For fully developed flow of a fluid of density 1000 kg/m³ and dynamic viscosity 0.001 Pa. s through a smooth pipe of diameter 10 mm with a velocity of 1 m/s, determine the Darcy friction factor________

Solution:

Given: density of the fluid, ρ =1000 kgm^{-3}

Dynamic viscosity of fluid μ = 0.001 Pa

Diameter of pipe, d = 10mm = 0.01 m

Velocity of fluid V = 1ms^{-1}

Reynolds number, Re $= \frac{\rho V d}{\mu} = \frac{1000\times 1\times 0.01}{0.001} = 10000$

Re = 10000 2100

Hence the flow is turbulent.

∴Darcy's friction factor, $f = \frac{0.3}{\text{Re}^{0.25}} = \frac{0.3}{(10000)^{0.25}} = 0.03$

Hence, the answer is 0.03.

Q16. Air flows steadily through a channel. The stagnation and static pressures at a point in the flow are measured by a Pitot tube and a wall pressure tap, respectively. The pressure difference is found to be 20 mm Hg. The densities of air, water and mercury, in kg/m³, are 1.18, 1000 and 13600, respectively. The gravitational acceleration is 9.81 m/s². Determine the air speed in m/s.____________

Solution:

Given:

Pressure difference between stagnation pressure and static pressure,

$p - p_s = 20$ mm Hg

$p - p_s = \rho_{Hg} \times gh$

$p - p_s = 13600\ 9.81\ 20\ 10^{-3}$ Pa

$p - p_s = 2668.32$ Pa

Now, we have $p = p_s + \frac{1}{2}\rho V^2$

where p = stagnation pressure

p_s = static pressure

ρ = density of air

$\therefore 2628.32 = \frac{1}{2} \times 1.18 \times V^2$

$V^2 = 4522.576$

$V = 67.25$ ms^{-1}

Hence, answer is 67.25 m/s.

Common Data Questions

Common Data for Questions 17 and 18:

The velocity field within a laminar boundary layer is given by the expression:

$$\vec{V} = \frac{BU_\infty y}{x^{3/2}}\hat{i} + \frac{BU_\infty y^2}{4x^{5/2}}\hat{j}$$

where $B = 100$ and the free-stream velocity $U_\infty = 0.1$ m/s.

Q17. Calculate the x-direction component of the acceleration in m/s² at the point $x = 0.5$ m and $y = 50$ mm.

Solution: ____

Given:

$$\vec{V} = \frac{BU_{\infty}y}{x^{3/2}}\hat{i} + \frac{BU_{\infty}y^2}{4x^{5/2}}\hat{j}$$

$$B = 100\text{m}^{1/2} \text{ and } U_{\infty} = 0.1\text{ ms}^{-1}$$

$$\text{Here } u = \frac{BU_{\infty}y}{x^{\frac{3}{2}}} \text{ and } v = \frac{BU_{\infty}y^2}{4x^{\frac{5}{2}}}$$

$$\text{Now } a_x = u\frac{\partial u}{\partial x} \text{ and } v\frac{\partial u}{\partial y} + \frac{\partial u}{\partial t}$$

$$a_x = \frac{BU_{\infty}y}{x^{\frac{3}{2}}} \times \left[BU_{\infty}y \times \left(-\frac{3}{2}\right)x^{-\frac{5}{2}}\right] + \frac{BU_{\infty}y^2}{4x^{\frac{5}{2}}} \times \frac{BU_{\infty}}{x^{\frac{3}{2}}} + 0$$

$$= -\frac{3}{2}\frac{(BU_{\infty}y)}{x^4} + \frac{(BU_{\infty}y)^2}{4x^4}$$

$$= \frac{(BU_{\infty}y)^2}{x^4}\left(-\frac{3}{2} + \frac{1}{4}\right)$$

$$= \frac{(BU_{\infty}y)^2}{x^4} \times \left(-\frac{5}{4}\right)$$

$$= \frac{(100 \times 0.1 \times 0.05)^2}{(0.5)^4} \times \left(-\frac{5}{4}\right)$$

$$= 4 \times \left(-\frac{5}{4}\right) = -5$$

Hence, the answer is -5 m/s^2.

Q18. Find the slope of the streamline passing through the point x= 0.5 m and y = 50 mm. ____

Solution: ____

$$\text{We have, } u = \frac{BU_{\infty}y}{x^{\frac{3}{2}}} \text{ and } v = \frac{BU_{\infty}y^2}{4x^{\frac{5}{2}}}$$

$$\text{Slope} = \frac{dy}{dx} = \frac{v}{u} = \frac{\dfrac{BU_{\infty}y^2}{4x^{\frac{5}{2}}}}{\dfrac{BU_{\infty}y}{x^{\frac{3}{2}}}} = \frac{y}{4x}$$

Slope at $x = 0.5$ m and $y = 50$ mm $= 0.05$m $= \dfrac{0.05}{4\times 0.5} = 0.025$

Hence, the answer is 0.025.

Common Data for Questions 19 and 20:

The wave and eddy resistance of a sea-going vessel, 96 m in length, driven at a velocity of 12 m/s, is to be determined. For this purpose, a 1/16 the scale model is employed in fresh water and the coefficient of resistance C_{we} of the model is found to be 1.47×10^{-4}. The quantity C_{we} is defined as $_{we}$/(/2), where $_{we}$ is the wave and eddy resistance, is the density, V is the velocity and L is the characteristic length. The density of sea water is 1026 kg/m^3.

Q19. The velocity in m/s, at which the model is towed, is

(A) 0.75 (B) 1.33 (C) 3 (D) 192

Solution:

Given:

Length of prototype, $L_p = 96$ m

Velocity of prototype, $V_p = 12$ ms^{-1}

Length of scale ratio, $\dfrac{L_m}{L_p} = \dfrac{1}{16}$

Coefficient of resistance of model, $(C_{we})_m = 1.47\ 10^{-4}$

Density of sea water, = 1026 kgm^{-3}

Now, gravity force is dominant for wave and eddy resistance, therefore Froude number for model and prototype will be sea

i.e., $Fr_m = Fr_p$

$$\frac{V_m}{\sqrt{L_m g}} = \frac{V_p}{\sqrt{L_p g}}$$

$$\frac{V_m}{V_p} = \frac{\sqrt{L_m}}{\sqrt{L_p}} = \frac{\sqrt{1}}{\sqrt{16}} = \frac{1}{4}$$

$$V_m = \frac{1}{4}V_p = \frac{1}{4}\times 12\,\text{m/s} = 3\ \text{ms}^{-1}$$

Hence, the answer is (C).

Q20. The resistance of the prototype, in kN, is

(A) 6 (B) 25

(C) 26.9 (D) 100.1

Solution:

We have, $L_p = 96\text{m}$ and $\dfrac{L_m}{L_p} = \dfrac{1}{16}$

$$\therefore \frac{L_m}{96} = \frac{1}{16}$$

$$L_m = \frac{96}{16}\text{m} = 6$$

$$\therefore L_m = 6 \text{ m}$$

Now, since the vessel is partially immersed is the sea water, viscous force is also dominant. Therefore, Reynolds number for model and prototype will be same, and since C_{we} is the function of Re_r it will be same for both model and prototype.

$$\therefore (C_{we})_p = 1.47\times10^{-4}$$

∴ Resistance of the prototype,

$$(F_{we})_p = \frac{1}{2}(C_{we})_p .\rho V_p^2 L_p^2$$

$$= \frac{1}{2}\times1.47\times10^{-4}\times1026\times12^2\times96^2 \text{ N}$$

$$= 100078.23 \text{ N}$$

$$= 100.07823$$

$$\approx 100.1\text{KN}$$

Hence, the answer is (D).

Linked Answer Questions

Statement for Linked Answer Questions 21 and 22:

Water enters a symmetric forked pipe and discharges into atmosphere through the two branches as shown in the Figure. The cross-sectional area of section-1 is 0.2 m² and the velocity across section-1 is 3 m/s. The density of water may be taken as 1000 kg/m³. The viscous effects and elevation changes may be neglected.

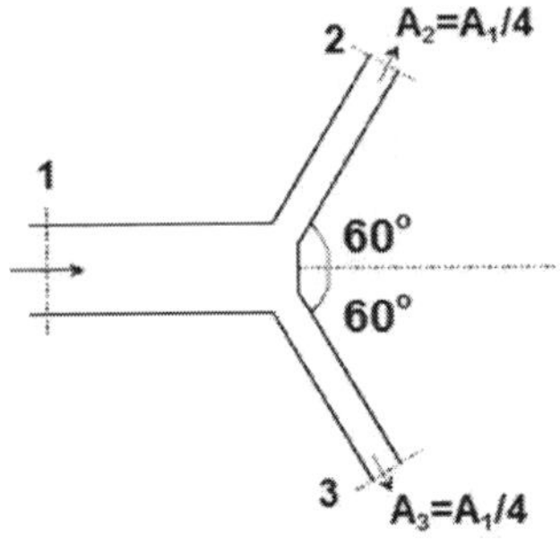

Q21. The gauge pressure at section-1, in kPa, is

(A) 0.6 (B) 13.5 (C) 135 (D) 600

Solution:

Given:

Cross sectional area for section, $A_1 = 0.2 \text{ m}^2$

Velocity across section, $V_1 = 3 \text{ ms}^{-1}$

Now, from continuity equation, we have,

$A_1V_1 = A_2V_2 + A_3V_3$

But, since section 2 and 3 are symmetric, therefore, $A_2 = A_3$ and $V_2 = V_3 = V$ (say)

$$\therefore A_1V_1 = \frac{A_1}{4}V + \frac{A_1}{4}V$$

$$A_1V_1 = \frac{A_1}{2}V$$

$$V_1 = \frac{V}{2}$$

$$V = 2V_1 = 2\times 2 \text{ m/s} = 6 \text{ ms}^{-1}$$

Now, applying Bernoulli's equation between section1and section 2, we have

$$\frac{p_1}{\rho g} + \frac{V_1^2}{2g} + Z_1 = \frac{p_2}{\rho g} + \frac{V_2^2}{2g} + Z_2$$

$$\frac{p_1}{\rho g} + \frac{3\times 3}{2g} = 0 + \frac{6\times 6}{2g} \quad (Z_1 = Z_2 \text{ and } p_2 = p_{atm} = 0 \text{ gauge pressure})$$

$$\frac{p_1}{\rho} + \frac{9}{2} = 18$$

$$\frac{p_1}{1000} = 18 - \frac{9}{2} = 13.5$$

$$p_1 = 13.5\times 10^3 \, Pa$$

$p_1 = 13.5$ kPa

Hence, the answer is (B).

Q22. The magnitude of the force, in kN, required to hold the pipe in place, is

(A) 2.7 (B) 5.4 (C) 19 (D) 27

Solution:

The magnitude of the force required to hold the pipe in place is given by

$F_x = Q_1V_1 + [Q_2V_2 X \cos(60°) + Q_3V_3\cos(60°)] + p_1A_1$

But due to symmetric, we have $Q_2 = Q_3$ and $V_2 = V_3$

$F_x = -\rho Q_1 V_1 + 2\rho Q_2 V_2 \cos(60°) + p_1 A_1$

$= -1000 \times 0.2 \times 3^2 + 2 \times 1000 \times \frac{0.2}{4} \times 6^2 \times \frac{1}{2} + 13500 \times 0.2 \text{ N}$

$= -1000 \times 0.2 \times 3^2 + 2 \times \frac{0.2}{4} \times 6^2 + 2700 \text{ N}$

$= 2700\ N = 2.7 \text{ kN}$

Hence, the answer is (A).

GATE 2012
FLUID MECHANICS

Q. 1 – Q. 9 carry one mark each.

Q1. In a two-dimensional flow field, the velocities of x and y directions are u and v, respectively. The shear stress for a Newtonian fluid having dynamic viscosity μ is given by-

(A) $\mu\left(\frac{\partial v}{\partial x}-\frac{\partial u}{\partial y}\right)$ (B) $2\mu\left(\frac{\partial v}{\partial y}\right)$ (C) $2\mu\left(\frac{\partial v}{\partial x}\right)$ (D) $\mu\left(\frac{\partial v}{\partial x}+\frac{\partial u}{\partial y}\right)$

Solution:

For two dimensional flow, angular deformation rate or rate of shear strain is given by

$$\dot{\gamma}=\left(\frac{\partial v}{\partial x}+\frac{\partial u}{\partial y}\right)$$

Now, shear stress is given by

$$\tau=\mu\dot{\gamma}=\mu\left(\frac{\partial v}{\partial x}+\frac{\partial u}{\partial y}\right)$$

Hence, the answer is (D).

Q2. In a potential flow, the superposition of the stream functions of a uniform flow and a line source give to a dividing streamline representing

(A) Rankine's half body (B) infinite circular cylinder

(C) Infinite rotating circular cylinder (D) infinite elliptical cylinder

Solution:

In a potential flow, the superposition of the stream function of uniform flow and a line source gives rises to a streamline representing Rankine's half body.

Hence, the answer is (A).

Q3. Given that V, L and g are the characteristic velocity, characteristic velocity and acceleration due to gravity respectively, the expression $\frac{V}{\sqrt{Lg}}$ represents

(A) Weber number (B) Euler number

(C) Cavitation number (D) Froude number

Solution:

Froude number is given by, $Fr = \frac{V}{\sqrt{Lg}}$

Hence, the answer is (D).

Q4. Match the devices in column I with the characteristics in column II

Column I	Column II
P. orifice meter	1. high head loss and low cost
Q. venturi meter	2. high head loss and high cost
	3. low head loss and high cost
	4. low head loss and low cost

(A) P-2; Q-4 (B) P-1; Q-2 (C) P-3; Q-1 (D) P-1; Q-3

Solution:

Orifice meter has high head loss but low cost, whereas venturi meter has low head loss but high cost.

Hence, the answer is (D).

Q5. Identify the visualization method that shows a PATHLINE in an unsteady flow, assuming that the camera covers the required field of view.

(A) A dye is continuously injected and a snapshot is taken

(B) A dye is continuously injected and a long exposure picture is taken

(C) A blob (or drop) of dye is injected and a snapshot is taken

(D) A blob (or drop) of dye is injected and a long exposure picture is taken.

Solution:

Pathline is the locus of path traced by a single fluid particle over a given period of time. Therefore, in order to observe the Pathline a drop of dye should be injected and a long exposure picture should be taken.

Hence, the answer is (D).

Q6. In the case of a fully developed flow through a pipe, the shear stress at the centerline is

(A) a function of the axial distance (B) a function of the centerline velocity

(C) zero (D) infinite

Solution:

For fully developed flow in a pipe, shear stress is given by

$$\tau = \left(\frac{dp}{dx}\right)\frac{r}{2}$$

Therefore at centerline, $r = 0$, therefore $\tau = 0$

Hence, the answer is (C).

Q7. The velocity u in 1D unsteady flow is given by (x^2-t), where x is the position and t is the time. The total acceleration at any x and t is

(A) $-1+xt+x^3$ (B) $-1+xt+2x^3$ (C) $-1-xt-x^3$ (D) $-1-2xt+2x^3$

Solution:

Given:

$$u = (x^2 - t)$$

Now, we have, $a_x = u\frac{\partial u}{\partial x} + v\frac{\partial u}{\partial x} + w\frac{\partial u}{\partial z} + \frac{\partial u}{\partial t}$

$$a_x = (x^2 - t) \times 2x + 0 + 0 - 1$$

$$a_x = 2x^3 - 2xt - 1$$

$$\therefore\ a_x = -1 - 2xt + 2x^3$$

Hence, the answer is (D).

Q8. If ψ is the stream function, Laplace equation $\nabla^2 \psi = 0$ is true when the flow is

(A) incompressible (B) incompressible and irrotational

(C) irrotational (D) compressible

Solution:

For a given stream function ψ, the Laplace's equation $\Delta^2 \psi = 0$ is true if the flow is incompressible irrotational.

Hence, the answer is (B).

Q9. A fully developed laminar flow is taking place through a pipe. If the flow velocity is doubled maintaining the flow laminar, the pressure loss would be

(A) halved (B) unaltered

(C) doubled (D) tripled

Solution:

For a fully developed laminar flow through a pipe, pressure loss it given by.

$$\Delta p = \frac{32\mu VL}{D^2},$$

where μ is the coefficient of dynamic viscosity, V is the average flow velocity, L is the length of the pipe and D is the diameter of the pipe.

Hence, $\Delta p \infty V$.

This means that the pressure loss will be doubled if the flow velocity is doubled.

Hence, the answer is (C).

Q. 10 - Q. 22 carry two marks each.

Q10. In the following equations u and v are the velocities in the x and y directions, respectively and t is time. The flow field that CANNOT be termed as incompressible is

(A) $u=x^3+xy^2$, $v=y^3+yx^2$ (B) $u=10xt$,v=-10yt

(C) $u=(y/^{1/7}$,v=0 () (D) $u=2y$, $v=2x$

Solution:

For the flow to be incompressible, it must satisfy the continuity equation.

(A) $u=x^3+xy^2$, $v=y^3+yx^2$

$$\therefore \frac{\partial u}{\partial x}+\frac{\partial v}{\partial y}=3x^2+y^2+3y^2+x^2=4\left(x^2+y^2\right)\neq 0$$

Hence, the flow is not incompressible.

(B) $u=10xt, v=-10yt$

$$\therefore \frac{\partial u}{\partial x}+\frac{\partial v}{\partial y}=10t-10t=0$$

Hence, the flow is incompressible.

(C) $u = (y/\delta)1/7$, $v = 0$

$$\therefore \frac{\partial u}{\partial x}+\frac{\partial v}{\partial y}=0+0=0$$

(D)Hence, the flow is incompressible.

$u=2y$, $v=2x$

$$\therefore \frac{\partial u}{\partial x}+\frac{\partial v}{\partial y}=0+0=0$$

Hence, the flow is incompressible.

Hence, the answer is (A).

Q11. A U-tube mercury manometer as shown below is employed to measure the pressure of an oil-filled vessel. The densities of Hg and oil are 13600 kg/m³ and 800 kg/m³ respect. The gravitational acceleration may be taken as 10 m/s². The gauge pressure (in Pa) at point A when h_1=0.5 m and h_2=0.9 m, is approximately

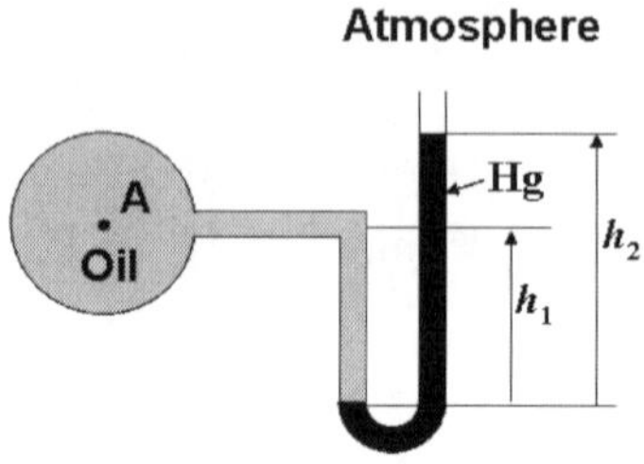

(A) 118.4×10^3 (B) 118.4
(C) 11.84 (D) 1.184

Solution:

Given:

Density of Hg, $\rho_{\text{Hg}} = 13600\,\text{kg/m}^3$

Density of oil, $\rho_{\text{oil}} = 800\,\text{kg/m}^3$

$h_1 = 0.5\text{ m and } h_2 = 0.9\text{ m}$

From the given figure, we have,

$$p_A + \rho_{\text{oil}} g h_1 = \rho_{\text{Hg}} g h_2$$

$$p_A = \rho_{\text{Hg}} g h_2 - \rho_{\text{oil}} g h_1$$

$$= (13600 \times 0.9 - 800 \times 0.5) \times 10\text{ Pa} = 118.4 \times 10^3\text{ Pa}$$

Hence, the answer is (A).

Q12. Water is supplied to a tank at the rate of 0.02 m³/s, as shown in figure below. The cross-sectional area of the tank is 1 m² and the inner diameter of the outlet pipe is 60 mm. At a time when the water level in the tank is increasing at the rate of 5 mm/s, the avg. velocity (in m/s) of water in the outlet pipe is approx.

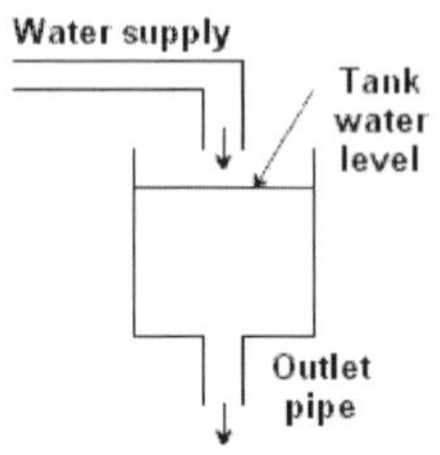

(A) 0.005 (B) 0.06 (C) 5.3 (D) 20

Solution:

Given:

Rate of water supplied, $Q_i = 0.2 \text{ m}^3 / \text{s}$

Cross sectional area of tank, $A = 1 \text{ m}^2$

Inner diameter of outlet pipe, $d = 60 \text{ mm} = 0.06 \text{ mm}$

Rate of increase of water level, in the tank, $\dfrac{dh}{dt} = 5 \text{ mm/s} = 0.005 \text{ m/s}$

Now, if Q_0 and Q are the flow rate in outlet pipe and flow rate at which water level rises in the tank respectively, then a/c to continually equation, we have

$$Q_i - Q_0 = Q$$

$$Q_0 = Q_i - Q$$

$$\frac{\pi d^2}{4} \times V = 0.02 - A \times \frac{dh}{dt}$$

where, V is the average velocity in outlet pipe

$$\therefore \frac{\pi \times (0.06)^2}{4} \times V = 0.02 - 1 \times 0.005$$

$$V = 5.3 \text{ m/s}$$

Hence, the answer is (C).

Q13. The water level in a gas pressurized tank with a large cross sectional area is maintained constant is shown in the figure below. The water level in the tank is 4.2 m above the pipe centerline as indicated in the figure. The gas pressure is 130 kPa. The atmospheric pressure, gravitational acceleration and density of water may be taken as 100 kPa, 10 m/s^2 and 1000 kg /m^3 respectively. Neglecting losses, the maximum velocity of water at any location in the horizontal portion of the delivery pipe for the pressure NOT to drop below atmospheric pressure is

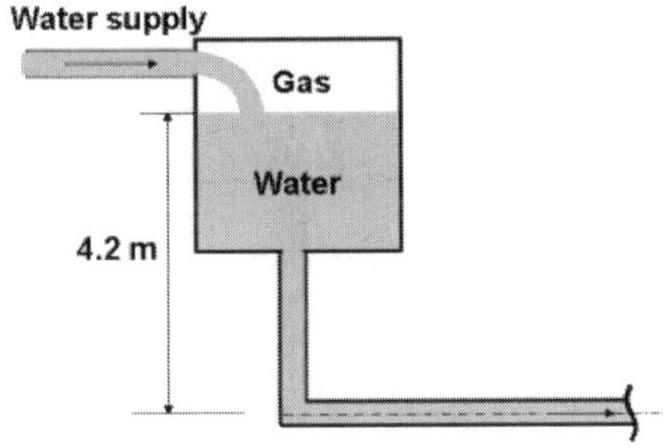

(A) 1.3 (B) 4.2 (C) 10 (D) 12

Solution:

Given:

Water level in the tank above the pipe centerline, $h_1 = 4.2\,\text{m}$

Gas pressure, $p = 130\,\text{kPa} = 1.3\times10^5\,\text{Pa}$

Hence the gauge pressure, $p_g = (130-100)\,\text{kPa} = 30\,\text{kPa} = 30\times10^3\,\text{Pa}$

Density of water, $\rho = 1000\,\text{kg/m}^3$

Now, we have $p_g = \rho g h_2$

where h_2 is the height of water above the water level in the tank which results a hydrostatic

pressure equal to p_g.

Therefore, $h_2 = \dfrac{p_g}{\rho g} = \dfrac{0.3\times10^5}{1000\times10}\,\text{m} = 3\,\text{m}$

Now, average velocity in the outlet pipe given by,

$V = \sqrt{2gh}$; where $h = h_1 + h_2 = (4.2+3)\,\text{m} = 7.2\,\text{m}$

$\therefore V = \sqrt{2\times10\times7.2}\,\text{m/s} = 12\,\text{m/s}$

Hence, the answer is (D).

Q14. The figure given below shows typical non-dimensional velocity profile for fully developed laminar flow between two infinitely long parallel plates separated by a distance a along y-direction. The upper plate is moving with a contant U in the x-direction and the lower plate is stationary.

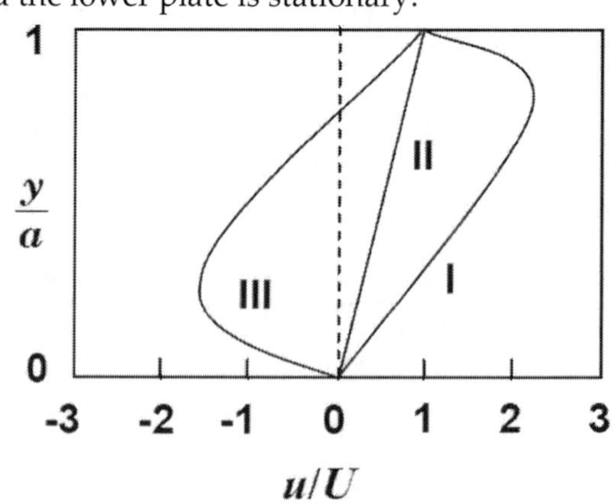

Match the non-dimensional velocity profiles in column I with the pressure gradients in column II

Column I		Column II
P. profile I	1.	$\frac{\partial p}{\partial x} > 0$
Q. profile II	2.	$\frac{\partial p}{\partial x} < 0$
R. profile III	3.	$\frac{\partial p}{\partial x} = 0$

(A) P-2; Q-3; R-1
(B) P-3; Q-2; R-1
(C) P-3; Q-1; R-3
(D) P-1; Q-2; R-3

Solution:

This is the case of a Couette flow. A Couette flow with a zero pressure gradient $\left(\text{i.e.,} \frac{\partial p}{\partial x} = 0\right)$ is called a simple Couette flow, for which the velocity increases linearly from the bottom plate to the upper plate. This means that curve II will correspond to the condition $\frac{\partial p}{\partial x} = 0$. Further, when $\frac{\partial p}{\partial x} < 0$ the situation is referred to as favourable pressure gradient. In such a case the velocity at some location in the gap between the two plates exceeds the top-plate velocity. This implies that curve I will correspond to the condition $\frac{\partial p}{\partial x} < 0$.. Likewise curve III will correspond to the condition $\frac{\partial p}{\partial x} > 0$.

Hence, the answer is (A).

Q15. Air flows over a spherical storage vessel of diameter 4 m at a speed of 1 m/s. To find the drag force on the vessel, a test run is to be carried out in water using a sphere of diameter 100 mm. The density and dynamic viscosity of air are 1.2 kg/m^3 and 1.8 ×10^{-5} Pa.s, respectively. The density and dynamic viscosity of water are 1000 kg/m^3 and 10^{-3} Pa. s, respectively. The drag force on the model is 4 N under dynamically similar conditions. The drag force (in N) on the prototype is approximately

(A) 0.25 (B) 0.93 (C) 1.08 (D) 4

Solution:

Given:

Diameter of spherical vessel, $d_p = 4$ m (prototype)

Speed of air, $V_p = 1$ m/s

Density of air, $\rho_p = 1.2$ kg/m$_3$

Viscosity of air, $\mu_p = 1.8 \times 10^{-5}$ Pa.s

Diameter of model, $d_m = 100$ mm = 0.1 m

Density of water, $\rho_m = 1000$ kg/m^3

Viscosity of water, $\mu_m = 10^{-3}$ Pa.s

Drag force on model $F_m = 4$ N

Now, drag force depends on coefficient of drag which depends on Reynolds number.

Therefore fore dynamic equilibrium between model and prototype, Reynolds number for model and prototype must be same.

i.e. $Re_m = Re_p$

$$\frac{\rho_m V_m d_m}{\mu_m} = \frac{\rho_p V_p d_p}{\mu_p}$$

$$\frac{1000 \times 0.1 \times V_m}{0.001} = \frac{1.2 \times 4 \times 1}{1.8 \times 10^{-5}}$$

$V_m = 2.66$ m/s

Now, we have $F_m = \frac{1}{2} C_D \rho_m A_m V_m^2$

$$4 = \frac{1}{2} C_D \times 1000 \times \frac{\pi \times (0.1)^2}{4} \times (2.66)^2$$

$$C_D = 0.144$$

Note: Drag coefficient for both model and prototype will be same, since it is a function of Re and Re for both model and prototype is the same.

$$\therefore F_p = \frac{1}{2} C_D \rho_p A_p V_p^2 = \frac{1}{2} \times 0.144 \times 1.2 \times \frac{\pi \times 4^2}{4} \times 1^2 1.08 \text{ N}$$

Hence, the answer is (C).

Q16. The velocity of an air stream is 20 m/s. The densities of mercury and air are 13600 kg/m^3 and 1.2 kg/m^3, respectively. The gravitational acceleration may be taken as 10 m/s^2. When a Pitot-static tube is placed in the stream, assuming the flow to be incompressible and frictionless, the difference between the stagnation and static pressure in the flow field (in mm Hg) would approximately be

(A) 1760 (B) 1.76 (C) 0.57 (D) 0.57×10^{-5}

Solution:

Given:

Velocity of air stream, $V = 20$ m/s

Density of mercury, $\rho Hg = 13600$ kg/m^3

Density of air, $\rho_{air} = 1.2$ kg/m^3

Now, stagnation pressure is equal to the sum of hydrostatic pressure and dynamic pressure i.e.

$$p = p_s + \frac{1}{2}\rho_{air}V^2$$

$$p - p_s = \frac{1}{2}\rho_{air}V^2$$

$$= \frac{1}{2}\times 1.2\times 20^2 \text{ Pa}$$

$$= 240 \text{ Pa}$$

If the difference between the stagnation and static pressures in terms of height of a mercury column is h, then

$$p - p_s = \rho_{Hg}gh$$

$$\therefore h = \frac{p - p_s}{\rho_{Hg}g} = \frac{240}{13600\times 10}\text{m} = 1.76\times 10^{-3}\text{m} = 1.76 \text{ mm}$$

Hence, the answer is (B).

Common Data for Questions 17 and 18:

A vessel containing water (density 1000 kg/m^3) and oil (density 800 kg/m^3), pressurized by gas, is shown in the figure below. Assume that the gravitational acceleration is 10 m/s^2.

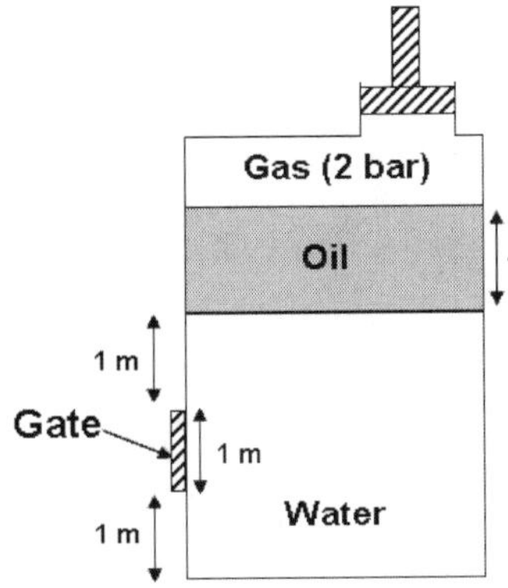

Q17. The pressure (in bar) exerted on the bottom wall inside the vessel is approximately

(A) 0.238 (B) 2.38 (C) 23.8 (D) 238

Solution:

The pressure exerted on the bottom wall inside the vessel is obtained as

$$p = p_{gas} + \rho_{oil} g h_{oil} + \rho_{water} g h_{water}$$
$$= 2\times10^5 + 800\times10\times1 + 1000\times10\times3 \text{ Pa}$$
$$= 2.38\times10^5 \text{ Pa}$$
$$= 2.38 \text{ bar}$$

Hence the answer is (B).

Q18. The gate is 1 m wide perpendicular to the plane of the paper. The force (in N) exerted on the gate is approx.

(A) 2.23×10^3 (B) 2.23×10^4 (C) 2.23×10^5 (D) 2.23×10^6

Solution:

The force exerted on the gate is given by

$$F = pA$$
$$= \left[p_{gas} + \rho_{oil} g h_{oil} + \rho_{water} g \left(h_{water}\right)_{gate}\right] A$$
$$= \left[2\times10^5 + 800\times10\times1 + 1000\times10\times1.5\right]\times1\times1 \text{ N}$$
$$= 223000 \text{ N}$$
$$= 2.23\times10^5 \text{ N}$$

Hence, the answer is (C).

Common data for questions 19 and 20

A boat is propelled in still water at a velocity of 5 m/s by taking water at the rate of $1 m^3/s$ from the aft side and discharging it through the stern using a pump as shown in the figure below. The velocity of the discharge jet relative to the boat is 9 m/s. The effect of pressure at the intake and discharge can be neglected. The density of water may be taken as $1000 kg/m^3$.

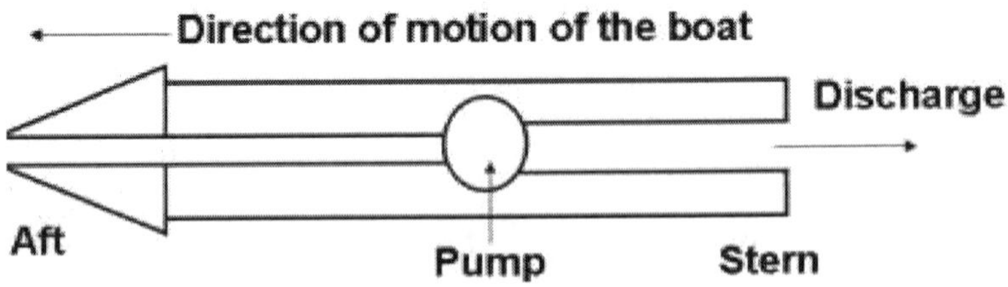

Q19. The power (in kW) reqd. to propel the boat is

(A) 10 (B) 20 (C) 50 (D) 90

Solution:

Velocity of boat, $V_B = 5$ m/s

Flow rate of water, $Q = 1 m^3/s$

Velocity of discharge rate relative to boat, $V_r = 9$ m/s

Density of water, $\rho_w = 1000$ kg/m^3

Now, reaction force exerted by set on the best, $F = \rho_w Q(V_r - V_B)$

∴ Power required to propel the boat is given by,

$P = F \times V_B = 4000 \times 5 = 20000$ W = 20 kW.

Hence, the answer is (B).

Q20. The total kinetic energy imparted to the water per second (in kW) by the pump is

(A) 10 (B) 25 (C) 28 (D) 81

Solution:

Total K.E imported to the water per second is given by,

$$P = \frac{1}{2}\rho Q\left(V_r^2 - V_B^2\right) = \frac{1}{2}\times 1000\times 1\times\left(9^2 - 5^2\right)W = 28000W = 28\text{kW}.$$

Hence, the answer is (C).

Linked Answer Questions

Statement for Linked Answer Questions 21 and 22:

The hydrodynamic boundary layer over a flat plate is shown in the figure below. The velocity in the x-direction is approximated as $u=a+by+cy^2$, where a, b and c are constants. U is the free stream velocity and is the boundary-layer thickness at any point x on the plate.

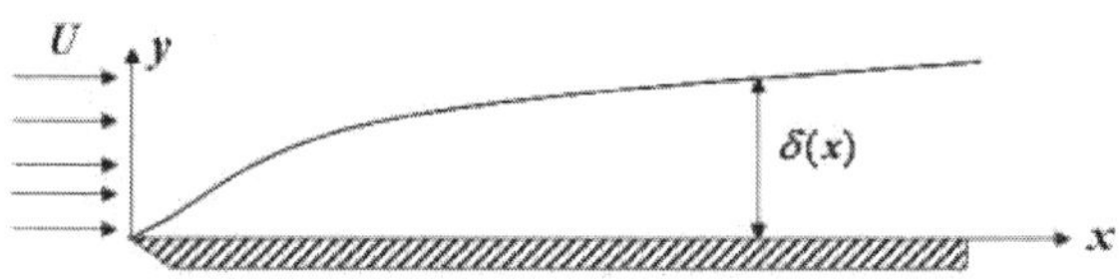

Q21. The dimensionless velocity profile is

(A) $\frac{u}{U} = 2\left(\frac{y}{\delta}\right) - \left(\frac{y}{\delta}\right)^2$ (B) $\frac{u}{U} = 2\left(\frac{y}{\delta}\right) + \left(\frac{y}{\delta}\right)^2$

(C) $\frac{u}{U} = 1.5\left(\frac{y}{\delta}\right) - 0.5\left(\frac{y}{\delta}\right)^2$ (D) $\frac{u}{U} = 1.5\left(\frac{y}{\delta}\right) + 0.5\left(\frac{y}{\delta}\right)^2$

Solution:

Given:

Velocity in the x- direction, $u = a + by + cy^2$.

Now, for a hydrodynamic boundary layer, we have three conditions: -

(i) At $y = 0$, $u = 0$

$\therefore 0 = a + b \times 0 + c \times 0^2$

$a = 0 \ldots$ (i)

(ii) At $y = \delta$, $u = U$

$\therefore u = a + by + cy^2$

$U = 0 + b\delta + c\delta^2 \ldots$ (ii)

(iii) At $y = d, \dfrac{du}{dy} = 0.$

Now, $\dfrac{du}{dy} = 0 + b + 2cy$

Hence $b + 2c\delta = 0$

$$cd = \frac{-b}{2} \qquad \ldots\text{(iii)}$$

Using eq. (iii) in eq. (ii) we get

$$U = 0 + bd + \left(\frac{-b}{2}\right) + \delta$$

$$U = \frac{b\delta}{2}$$

$$b = \frac{2U}{\delta}$$

$$\therefore c\delta = \frac{-b}{2} = \frac{-2U}{2\delta}$$

$$c = \frac{-u}{\delta^2}$$

Putting the value of a, b and c in the given velocity profile, we have,

$$u = 0 + \left(\frac{2U}{\delta}\right)y + \left(\frac{-U}{\delta^2}\right)y^2$$

$$\frac{u}{U} = 2\left(\frac{y}{\delta}\right) - \left(\frac{y}{\delta}\right)^2$$

Hence, the answer is (A).

Q22. The displacement thickness (in mm) when= 6 mm, is

(A) 2.25 (B) 2 (C) −2 (D) −2.25

Solution:

We have

$$\frac{u}{U} = 2\left(\frac{y}{\delta}\right) - \left(\frac{y}{\delta}\right)^2$$

Now, displacement thickness is given by,

$$\theta^* = \int_0^\delta \left(1 - \frac{u}{U}\right) dy$$

$$= \int_0^\delta \left(1 - \frac{2y}{\delta} + \frac{y^2}{\delta^2}\right) dy$$

$$= \int_0^\delta dy - \frac{1}{\delta}\int_0^\delta 2y dy + \frac{1}{\delta^2}\int_0^\delta y^2 dy$$

$$= [y]_0^\delta - \frac{2}{\delta}\left[\frac{y^2}{2}\right]_0^\delta + \frac{1}{\delta^2}\left[\frac{y^3}{3}\right]_0^\delta$$

$$\theta^* = \delta - \delta + \frac{\delta}{3}$$

$$\theta^* = \frac{\delta}{3} = \frac{6}{3}\text{ mm} = 2\text{ mm}$$

Hence, the answer is (B).

GATE 2011
FLUID MECHANICS

Q. 1 - Q. 9 carry one mark each.

Q1. For a boundary layer on a flat plate _______ forces and ________ forces are of the same order of magnitude.

(A) body, inertia
(B) viscous, body
(C) inertia, viscous
(D) viscous, pressure

Solution:

For a boundary layer on a flat plate inertia and viscous forces are of the same order of magnitude.

Hence, the answer is (C).

Q2. The temperature field in a fluid flow is given by (60 – 0.2xy)°C. The velocity field is $\vec{V} = 2xy\hat{i} + ty\hat{j}$ m/s. The rate of change of the temperature measured by a thermometer moving along with the flow at (2, -4) m at t = 4 s is

(A) -12.8°C/s
(B)-10.6°C/s
(C) -6.4 °C/s
(D) -4.8 °C/s

Solution:

Given:

Temperature, T = (60 – 0.2xy)°C.

Velocity, $\vec{V} = 2xy\hat{i} + ty\hat{j}$.

We here to find the rate of change of temperature at (2, -4) and t = 4 s.

Now, we have $\frac{dT}{dt} = \frac{\partial T}{\partial x}\frac{dx}{dt} + \frac{\partial T}{\partial y}\frac{dy}{dt} + \frac{\partial T}{\partial t}$

or $\frac{dT}{dt} = u\frac{\partial T}{\partial x} + v\frac{\partial T}{\partial y} + \frac{\partial T}{\partial t}$

Here, $u = 2xy$ and $v = ty$.

Hence $\frac{dT}{dt} = 2xy \times (0 - 0.2y) + ty \times (0 - 0.2x) + 0$

$$\frac{dT}{dt} = -0.4xy^2 - 0.2xyt$$

Therefore, $\left.\frac{dT}{dt}\right]_{(x=2, y=-4, t=4)} = -0.4 \times 2 \times (-4)^2 - 0.2 \times 2 \times (-4) \times 4 = -6.4°\text{C.s}^{-1}$

Hence, the answer is (C).

Q3. Two tanks, A and B, with the same height are filled with water till the top. The volume of tank A is 10 times the volume of tank B. What can you say about the pressures p_A and p_B at the bottom of the tanks A and B respectively?

A) $p_A = 10 p_B$

(B) $p_B = 10 p_A$

(C) $p_A = p_B$

(D) Additional data is required to compare the two pressures.

Solution:

Water is filled up to the same height in both the tank A and B.

i.e, $h_A = h_B$

$\therefore p_A = \rho g h_A$ and $p_B = \rho g h_B$

$\therefore p_A = p_B$

Hence, the answer is (C).

Q4. A velocity field in a plane flow is given by $\vec{V} = 2xy\hat{i} + 3y\hat{j}$ m/s. The vorticity at the point (2, 4) m is

(A) $-4\hat{k}$ rad/s (B) $-3\hat{j}$ rad/s

(C) $-2\hat{k}$ rad/s (D) $-3\hat{i}$ rad/s

Solution:

Given: $u = 2xy$ and $v = 3y$

Now, vorticity is given by,

$$\Omega = \left(\frac{\partial v}{\partial x} - \frac{\partial u}{\partial y} \right) \hat{k}$$

$$\Omega = (0 - 2x)\hat{k}$$

$$\therefore \Omega \text{ at } (2,4) - 2 \times 2\hat{k} = -4\hat{k} \text{ rad/s,}$$

Hence, the answer is (A).

Q5. Separation is said to occur at a wall when ________ at the wall becomes zero.

(A) internal energy (B) pressure

(C) shear stress (D) density

Solution:

At the point of separation of boundary layer, shear stress becomes zero.

Hence, the answer is (C).

Q6. A certain fluid flow is influenced by density (ρ), angular velocity (), dynamic viscosity (), and a characteristic length (L). A relevant non-dimensional parameter will be

(A) $\dfrac{\rho\omega\mu}{L^2}$ (B) $\dfrac{\rho\omega L^2}{\mu}$

(C) $\rho\omega\mu L^2$ (D) $\rho\omega\mu L$

Solution:

Out of the four given options, (B) is nothing but the Reynold number,

i.e., $\text{Re} = \dfrac{\rho VL}{\mu} = \dfrac{\rho(\omega L)L}{\mu} = \dfrac{\rho\omega L^2}{\mu}$,

which is dimensionless

Hence, the answer is (B).

Q7. The drags due to potential flow past a cylinder of diameter *D* cm and a slender airfoil of chord length *D* cm are compared. Assuming unit depth for both the bodies, which one of the following would be TRUE?

(A) The drag on the cylinder is greater

(B) The drag on the airfoil is greater

(C) Both the drags are equal

(D) Additional data is needed to compare the drags

Solution:

For potential flow over a symmetrical body, the pressure distribution is symmetrical therefore form drag is zero. Moreover, since potential flow is inviscid, hence there is no friction. Accordingly there the friction drag has to be zero. Therefore, the drag forces on both cylinder and airfoil are zero.

Hence, the answer is (C).

Q8. For a fully developed flow between two parallel flat plates, the velocity gradient at a point is found to be 1000 s^{-1}. If the density of the fluid is 880 kg/m^3 and the kinematic viscosity of the fluid is 7.4 x 10^{-7} m^2/s, the shear stress at the same point is approximately

(A) 0 Pa

(B) 1.30 Pa

(C) 0.32 Pa

(D) 0.65 Pa

Solution:

Given: $\frac{du}{dy} = 1000\,\text{s}^{-1}$

Density $\rho = 880\,\text{kg}/\text{m}^3$

Kinetic viscosity, $\nu = 7.4\times10^{-7}\ \text{m}^2/\text{s}$

∴ Dynamic viscosity,

$$\mu = \rho\nu = 880\times7.4\times10^{-7}\ \text{N.s.m}^{-2} = 0.6512\times10^{-3}\ \text{N.s.m}^{-2}$$

Now, shear stress is given by,

$$\tau = \mu\frac{du}{dy} = 0.6512\times10^{-3}\times1000\ \text{Nm}^{-2} = 0.6512\ \text{Nm}^{-2}$$

Hence, the answer is (D).

Q9. An open channel flow is to be simulated in the laboratory. For this purpose, a 1:25 scale model is constructed. If the flow velocity in the prototype is 5 m/s, for dynamic similarity the model should have a flow velocity of

(A) 5 m/s

(B) 1 m/s

(C) 125 m/s

(D) 25 m/s

Solution:

Given:

Model scale $\frac{L_m}{L_p} = \frac{1}{25}$

Velocity of prototype, $V_p = 5\,\text{m}/\text{s}$

For dynamic similarity in open channel, Froude number for model and for protype must be same

$$\text{Fr}_m = \text{Fr}_p$$

$$\frac{V_m}{\sqrt{L_m g}} = \frac{V_p}{\sqrt{L_p g}}$$

i.e. $$\frac{V_m^2}{L_m} = \frac{V_p^2}{L_p}$$

$$\frac{V_m^2}{V_p^2} = \frac{L_m}{L_p}$$

$$\frac{V_m^2}{5^2} = \frac{1}{25}$$

$$V_m^2 = 1$$

$$V_m = 1$$

Hence, the answer is (B).

Q. 10 - Q. 22 carry two marks each.

Q10. A pilot-static probe is inserted in an air flow. A manometer connected to this probe having Hg as the manometric fluid shows a difference of 30 mm. Assume a probe factor of 1. Assuming ϱ_{air} = 1.23 kg/m^3, ϱ_{Hg} = 13600 kg/m^3 and g = 10 m/s^2, the speed of the air flow is approximately

(A) 66.5 m/s (B) 81.5 m/s

(C) 76.5 m/s (D) 92.5 m/s

Solution:

Given:

Monometric reading, x = 30 mm = 0.03 m

Density of air, ρ_{air} = 1.23 kg/m^3

Density of mercury, ρ_{Hg} = 13600 kg/m^3

Now velocity of fluid measured by pitot-tube is given by

$$V = \sqrt{2gx\left(\frac{\rho_{\text{Hg}}}{\rho_{\text{air}}} - 1\right)}$$

$$= \sqrt{2 \times 10 \times 0.03 \times \left(\frac{13600}{1.23} - 1\right)}\ \text{m/s}$$

$$= 81.5\ \text{m/s}$$

Hence, the answer is (B).

Q11. Consider an L-shaped gate with water level above the hinge as shown. At approximately what height D of the water level will the gate open? Neglect the mass of the gate. Assume $g = 10\ m/s^2$.

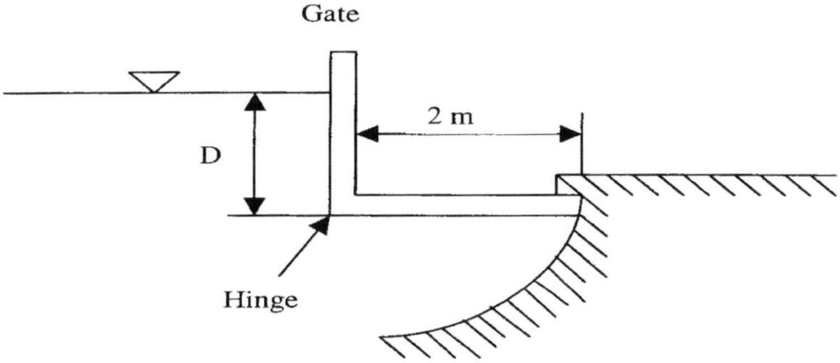

(A) 3.46m (B) 4.36

(C) 6.43m (D) 5.36m

Solution:

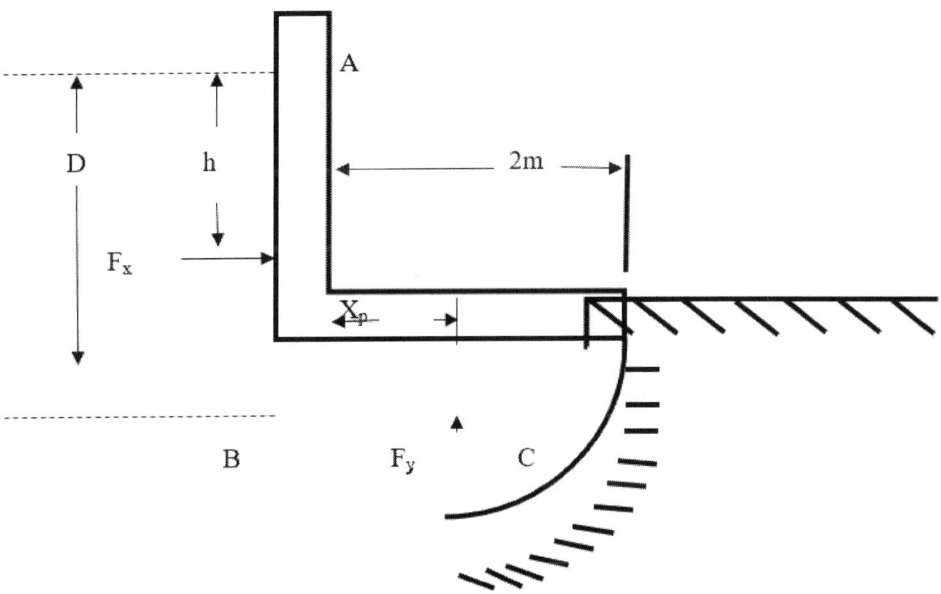

Let the depth of the gate into the plane of paper be 1 unit.

Now, the horizontal force exerted by the water on the gate is given by,

$$F_x = \rho g\bar{h}A,$$

where

ρ = density of water

$\bar{h}$ = depth of the centroid from the free surface of water

A = surface area of gate facing the water

$$\therefore \bar{h} = \frac{D}{2}\,\text{m}$$

and $A = D \times 1\text{ m} = D\text{ m}$

$$\therefore F_x = \rho g \frac{D}{2} D = \rho g \frac{D^2}{2} \qquad \text{...(i)}$$

Now, position of centre of pressure is given by,

$$h^* = \frac{I_G}{A\bar{h}} + \bar{h},$$

where I_G = Moment of area of the gate about an axis passing through centroid of the gate and parallel to the free surface.

$$\text{Now, } I_G = \frac{1}{12} \times 1 \times D^3$$

$$\therefore h^* = \frac{\frac{1}{12} \times 1 \times D^3}{D \times 1 \times \frac{D}{2}} + \frac{D}{2} = \frac{D}{6} + \frac{D}{2} = \frac{2}{3}D$$

Now the vertical force exerted by the water on the horizontal portion of the gate is given by;

F_y = the weight of water supported by the gate up to free surface

i.e., $F_y = \rho g A_h D$

where $A_h = 2 \times 1\text{ m}^2 = 2\text{ m}^2$

Hence, $F_y = \rho g \times 2 \times D = 2\rho g D$

The vertical force acts along the line passing through centre of gravity of the water supported by the horizontal portion of gate, i.e $x_p = \frac{2}{2}\text{ m} = 1\text{ m}$

Now, for the gate to be opened, the moment of F_x about point B must be equal to that of F_y about point B.

i.e., $F_x \times (D - h^*) = F_y \times x_p$

$$\Rightarrow \rho g \frac{D^2}{2} \times \left(D - \frac{2}{3}D\right) = 2\rho g D \times 1$$

$$\Rightarrow \rho g \frac{D^2}{2} \times \frac{D}{3} = 2 \rho g D$$

$$\Rightarrow D^2 = 12$$

$$\Rightarrow D = 3.46$$

Hence, the answer is (A).

Q12. When a large tank containing water is placed on a weighing scale, a reading of 10000 N is obtained. The tank is fitted with an outlet pipe and a valve as shown. When the valve is opened, a jet of water with a velocity of 10 m/s issues out in the vertically upward direction. The diameter of the outlet pipe is 10 cm. Determine approximately the reading on the weighing scale at the instant the valve is opened and the water jet issues out. Density of water is 1000 kg/m³.

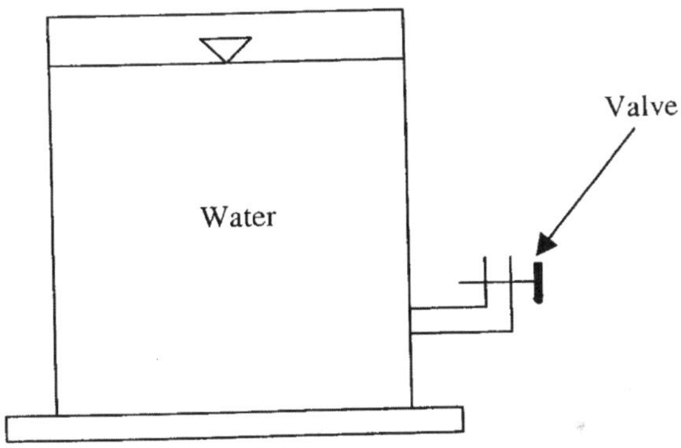

(A) 9215N (B) 10000 N (C) 10785 N (D) 12500 N

Solution:

Given:

Velocity of jet, $V = 10 \text{ m/s}$

Diameter of outlet pipe, $D = 10 \text{ cm} = 0.1 \text{ m}$

Weight of water, W = 10,000 N.

Now, when the value is opened, the reaction force exerted by the jet in downward direction is given by,

$$F_r = \rho Q V$$

$$= \rho \times \frac{\pi D^2}{4} \times V \times V$$

$$= \frac{1000 \times \pi \times 0.1^2 \times 10^2}{4} \text{ N}$$

$$= 785 \text{ N}$$

Therefore net force in downward direction,

$$F = W + F_r = (10000 + 785)\ \text{N} = 10785\ \text{N}.$$

Hence, the answer is (C).

Q13. A fluid with a volumetric flow rate of 5 m³/s enters the nozzle shown below. The cross-sectional area varies with x as $A(x) = 1/(1 + x^2)$. Assuming that the flow is parallel and uniform at each cross-section, the acceleration at any point in the nozzle is given by

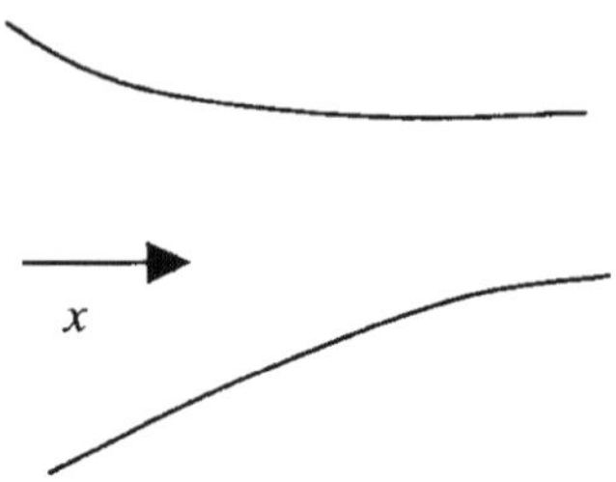

(A) 50 $(x + x^3)$ (B) 50 $(1 + x^2)$ (C) 0 (D) 50 $(x^2 + x^3)$

Solution:

Given:

Volumetric flow rate, Q = 5 m³/s

Cross – sectional area, $A(x) = \dfrac{1}{1+x^2}$

Therefore, average velocity or any point x is given by,

$$u(x) = \frac{Q}{A(x)} = \frac{5}{\{1/(1+x^2)\}} = 5(1+x^2)$$

Now, acceleration in x- direction is given by

$$a_x = u\frac{du}{dx} = 5(1+x^2)\times(5\times 2x) = 5(1+x^2)\times 10x = 50(x+x^3)$$

Hence, the answer is (A).

Q14. Consider fully developed flow of water in a pipe of diameter 2 cm. The average velocity of the flow is 2 m/s. The viscosity of the water is 10^{-3} kg/m-s and the density is 1000 kg/m³. The friction factor can be calculated using $f = 64 / \text{Re}$ for laminar flows and $f = 0.3164 / \text{Re}^{025}$ for turbulent flows. The pressure drop over a length of 0.5 m is

(A) 0.08 Pa (B) 325 Pa

(C) 1115 Pa (D) 9875 Pa

Solution:

Given:

Diameter of pipe, $D = 2\text{ cm} = 0.02\text{ m}$

Average velocity, $V = 2\text{ m/s}$

Length, $L = 0.5\text{ m}$

Now, Reynold's number is given by, $\text{Re} = \dfrac{\rho VD}{\mu} = \dfrac{1000 \times 2 \times 0.02}{10^{-3}} = 40000$

Since, $\text{Re} > 2100$, hence the flow is turbulent.

Therefore, $f = \dfrac{0.3164}{\text{Re}^{0.25}} = \dfrac{0.3164}{(40,000)^{0.25}} = 0.0223$

Now, pressure drop is given by,

$$\Delta p = f \frac{L}{d} \frac{\rho v^2}{2} = \frac{0.0223 \times 0.5 \times 1000 \times 2 \times 2}{0.02 \times 2} \text{ Pa} = 1115 \text{ Pa}$$

Hence, the answer is (C).

Q15. Consider a steady, fully developed flow in a horizontal pipe of diameter D. Over a section of length L of this pipe, a pressure drop of Δp is observed. The average wall shear stress over this section is

(A) $\dfrac{\Delta pD}{4L}$ (B) $\dfrac{\Delta pD}{2L}$ (C) $\dfrac{\Delta p\pi L}{2D}$ (D) $\dfrac{\Delta p\pi L}{4D}$

Solution:

For steady fully developed flow in a horizontal pipe, shear stress is given by

$$\tau = \frac{\Delta p}{L} \frac{r}{2}$$

Where r is the radial distance from the centre line of the pipe

Therefore, the shear stress on the wall is given by,

$$\tau_{wall} = \frac{\Delta p}{L} \times \frac{r}{2} = \frac{\Delta p}{L} \times \frac{D}{2 \times 2} = \frac{\Delta pD}{4L}$$

Hence, the answer is (A).

Q16. In an inviscid incompressible flow, the velocity field is given by $\vec{V} = x\hat{i} + y\hat{j}$ m/s and the body force per unit mass is given by $\vec{g} = -10\vec{k}$ m/s^2. The pressure at the point (0, 0, 0) is 101 Pa. Assuming that the density of the fluid is 1 kg/m^3, the pressure at the point (1, 1, 1) for this flow is

(A) 100 Pa (B) 105 Pa (C) 95 Pa (D) 90 Pa

Solution:

Given:

Velocity field, $\vec{V} = x\hat{i} + y\hat{j}$. i.e., $u = x$ and $v = y$

Pressure at the point (0, 0, 0), $p_1 = 101$ Pa

Velocity at the point (0, 0, 0), $\vec{V}_1 = 0\times\hat{i} + 0\times\hat{j}$

Hence, speed at the point (0, 0, 0), $V_1 = \sqrt{0^2 + 0^2} = 0$

Velocity at the point (1, 1, 1), $\vec{V}_2 = 1\times\hat{i} + 1\times\hat{j}$

Hence, speed at the point (1, 1, 1), $V_2 = \sqrt{1^2 + 1^2} = \sqrt{2}$ m/s

Elevation of point (0, 0, 0), $z_1 = 0$ m
Elevation of point (1, 1, 1), $z_2 = 1$ mm
Now applying the Bernoulli's equation between the points (0, 0, 0) and (1, 1, 1)

$$\frac{p_1}{\rho g} + \frac{V_1^2}{2g} + z_1 = \frac{p_2}{\rho g} + \frac{V_2^2}{2g} + z_2$$

$$\Rightarrow \frac{101}{1\times 10} + 0 + 0 = \frac{p_2}{1\times 10} + \frac{\left(\sqrt{2}\right)^2}{2\times 10} + 1$$

$$\Rightarrow \frac{p_2}{10} = \frac{101}{10} - \frac{1}{10} - 1$$

$$\Rightarrow p_2 = \left(101 - 1 - 10\right) \text{Pa}$$

$$\Rightarrow p_2 = 90 \text{ Pa}$$

Hence, the answer is (D).

Common Data Questions

Common Data for Questions 17 and 18:

A two-dimensional rectangular water jet of velocity 10 m/s and area 5 cm² impinges normal to a flat plate and splits symmetrically into two half jets, each of area 2.5 cm² as shown. Assume steady flow and neglect viscous effects and the weight of the plate and the water. Density of water is 1000 kg/m³.

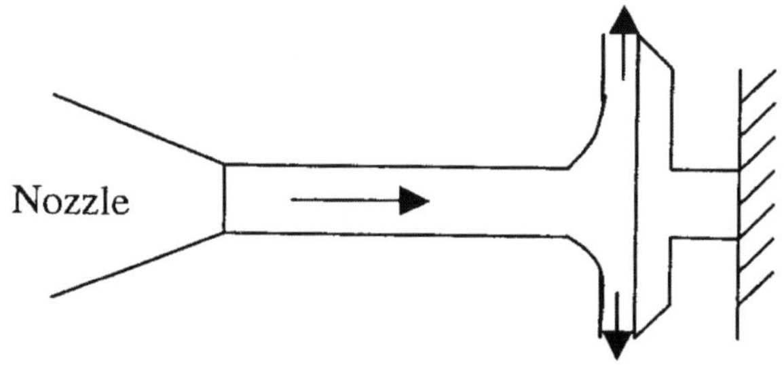

Q17. After splitting, the velocity of the upward half-jet along the plate is

(A) 5 m/s (B) 7.5 m/s (C) 2.5 m/s (D) 10 m/s

Solution:

Given:

Velocity of the water jet, $V = 10$ m/s

Area of the jet, $A = 5\ \text{cm}^2 = 5 \times 10^{-4}\ \text{m}^2$

Area of each jet after splitting, $A_1 = A_2 = 2.5\ \text{cm}^2 = 2.5 \times 10^{-4}\ \text{m}^2$

Density of water, $\rho = 1000\ \text{kg/m}^3$

From continuity equation, we have

$$Q = Q_1 + Q_2$$
$$\Rightarrow AV = A_1V_1 + A_2V_2$$

However, the set splits symmetrically. Therefore, $V_1 = V_2$

$$\therefore 5\times10^{-4}\times10 = 2.5\times10^{-4}\times V_1 + 2.5\times10^{-4}\times V_2$$
$$\Rightarrow 2.5\times10^{-4}\times(V_1+V_2) = 5\times10^{-4}\times10$$
$$\Rightarrow 2.5\times10^{-4}\times2V_1 = 5\times10^{-4}\times10$$
$$\Rightarrow V_1 = 10\ \text{m/s}$$

Hence, the answer is (D).

Q18. The magnitude of the reaction force at the wall is

(A) 20N (B) 25N (C) 35 N (D) 50 N

Solution:

The magnitude of the reaction force is given by

$$F_x = \rho Q(V_{1x} - V_{2x})$$
$$= 1000\times10\times5\times10^{-4}\times(10-0)$$
$$= 50\ \text{N}$$

Hence, the answer is (D).

Common Data for Questions 19 and 20:

A flow has a velocity field given by $\vec{V} = 2x\hat{i} - 2y\hat{j}$

Q19. The velocity potential $\varnothing(x, y)$ for the flow is

(A) $2x - 2y$ + const. (B) $2xy$ + const.

(C) $x^2 + y^2$ + const. (D) $x^2 - y^2$ + const.

Solution:

Given:

Velocity field, $\vec{V} = 2x\hat{i} - 2y\hat{j}$, i.e. $u = 2x$ and $v = -2y$

Now, for a given velocity potential $\varphi(x, y)$ we have,

$$u = \frac{\partial\varphi}{\partial x} \text{ and } v = \frac{\partial\varphi}{\partial y}$$

$$\therefore 2x = \frac{\partial\varphi}{\partial x} \text{ and } -2y = \frac{\partial\varphi}{\partial y}$$

$$\int d\varphi = \int 2xdx + f_1(y) \text{ or} \int d\varphi = -\int 2ydy + f_2(x)$$

$$\varphi = x^2 + f_1(y) \text{ or } \varphi = -y^2 + f_2(x)$$

$$\varphi = x^2 + f_1(y) \qquad ...(i)$$

$$\varphi = x^2 + f_1(y) \qquad ...(ii)$$

Therefore from eq. (i) and eq. (ii), we get

$$\varphi = x^2 - y^2$$

Hence, the answer is (D).

Q20. The streamlines for the velocity field look like

(A) (B) (C) (D)

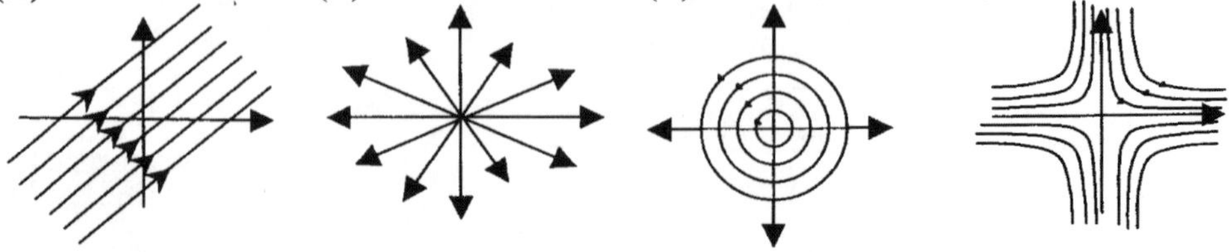

Solution:

We have $u = 2x$ and $v = -2y$.

Now, for a given stream function $\psi(x, y)$, we have,

$$u = \frac{\partial \psi}{\partial y} = 2x \ and \ v = -\frac{\partial \psi}{\partial x} = -\ 2y$$

Therefore,

$$\frac{\partial \psi}{\partial y} = 2x \quad \text{...(i)}$$

and

$$\frac{\partial \psi}{\partial x} = 2y \quad \text{...(ii)}$$

From eq. (i) and eq. (ii) we obtain

$\psi = 2xy + c.$

∴ Equation of streamline is given by, $\psi = 2xy + c.$

xy = constant = k (say)

$$x = \frac{k}{y} \quad \text{...(iii)}$$

Therefore, the graph for equation (iii) is represented in option (D).

Hence, the answer is (D).

Linked Answer Questions

Statement for Linked Answer Questions 21 and 22:

Two flat parallel plates are separated by a small gap h filled with an incompressible fluid of viscosity μ. Assume that the length and width of the plates to be much larger than the gap h. The top plate moves horizontally while the bottom plate is held stationary. The magnitude of the difference between the shear stress at the top and bottom walls is found to be $\Delta\tau$.

Q21. The velocity of the top plate is

(A) $\frac{h\Delta\tau}{2\mu}$ (B) $\frac{h\Delta\tau}{2\mu}$

(C) $2\frac{h\Delta\tau}{\mu}$ (D) $3\frac{h\Delta\tau}{2\mu}$

Solution:

This is the Couette the flow, where shear stress is given by:

$$\tau_{\text{top}} = \frac{\mu U_\infty}{h}(1+\alpha);\ \alpha = \frac{h^2}{2\mu U_\infty}\left(\frac{-dp}{dx}\right);\ \alpha \text{ is dimensionless pressure gradient}$$

$$\tau_{\text{bottom}} = -\frac{\mu U_\infty}{h}(1+\alpha);$$

For simple Couette flow, $\alpha = 0$.

$$\therefore \Delta\tau = \frac{2\mu U_\infty}{h}$$

$$U_\infty = \frac{h\Delta\tau}{2\mu}$$

Hence, the answer is (A).

Q22. If a finite width slender object is introduced parallel to the plates in the middle of the gap, the time at which it would have rotated clockwise by 90° would be

(A) $2\pi\,\mu/\Delta\tau$ (B) $\pi\mu/(\Delta\tau)$ (C) $2\pi\mu/(3\Delta\tau)$ (D) $\pi\mu/(4\Delta\tau)$

Solution:

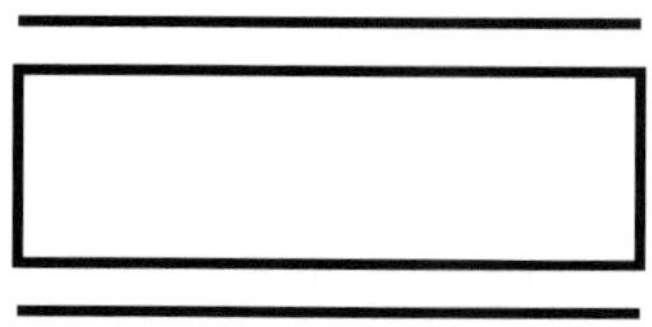

We have, $U_\infty = \dfrac{h\Delta\tau}{2\mu}$

Therefore, the angular velocity of the slender object, $\omega = \dfrac{U_\infty}{h} = \dfrac{h\Delta\tau}{2\mu h} = \dfrac{\Delta\tau}{2\mu}$

And angular displacement, $\theta = \dfrac{\pi}{2}$

Now, we have $\theta = \omega \times t$

$$\frac{\pi}{2} = \frac{\Delta\tau}{2\mu} \times t$$

$$t = \frac{\pi\mu}{\Delta\tau}$$

Hence, the answer is (B).

GATE 2010
FLUID MECHANICS

Q1. A cylindrical container is filled up to half of its height. The container is mounted on the centre of turn table and is fixed using a spindle .The turn table is now rotated about its central axis width a certain angular velocity. After some time interval, fluid attains rigid body rotation. Which of following profiles best represents the constant pressure surface on the container?

(A) 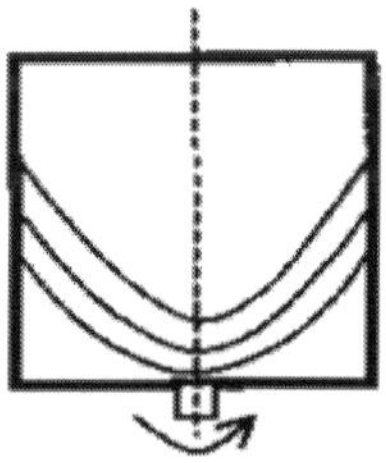

(B) 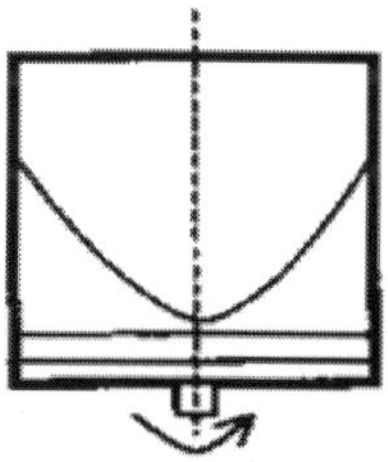

(C) 

(D) 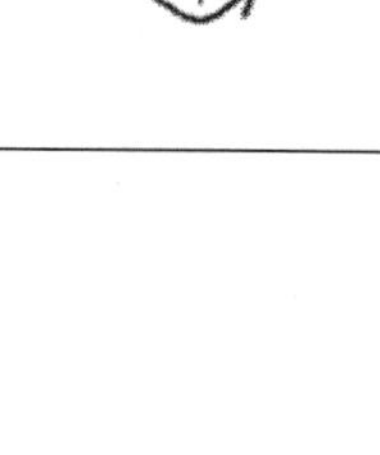

Solution: ____________________

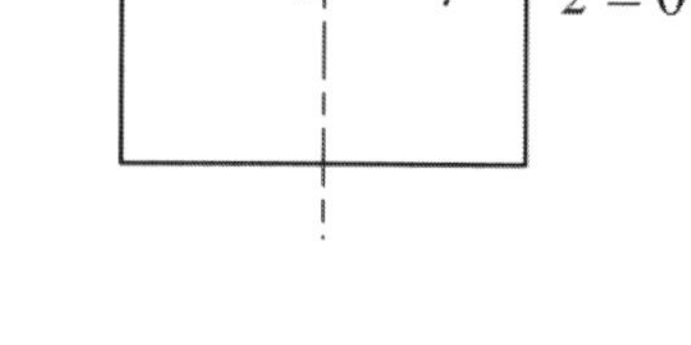

This is the case of a forced vector flow, where the velocity profile is parabolic. For forced vortex flow, pressure distribution is given by

$$p - p_{atm} = \frac{\rho\omega^2 r^2}{2} - \rho g z \qquad \text{...(i)}$$

Therefore the pressures p_1 and p_2 at two points with coordinates (r_1, z_1) and (r_2, z_2), respectively, are given by

$$p_1 - p_{atm} = \frac{\rho\omega^2 r_1^2}{2} - \rho g z_1 \qquad \text{...(ii)}$$

$$p_2 - p_{atm} = \frac{\rho\omega^2 r_2^2}{2} - \rho g z_2 \qquad \text{...(iii)}$$

If the pressures p_1 and p_2 are equal then from eqs. (ii)-(iii) one obtains

$$\frac{\rho\omega^2 r_1^2}{2} - \rho g z_1 = \frac{\rho\omega^2 r_2^2}{2} - \rho g z_2$$

$$\Rightarrow \frac{\rho\omega^2 r^2}{2} - \rho g z = \text{constant} \qquad \text{...(iv)}$$

$$\Rightarrow z = \frac{\omega^2 r^2}{2g} + \text{constant},$$

which is the equation of a parabola.

Hence, the answer is (A).

Q2. Match the items given in the following two columns using appropriate combinations:

	Column 1		Column 2
P	Ratio of inertial force to viscous force	1.	Reynolds number (Re)
Q	Ratio of momentum diffusivity and thermal diffusivity	2.	Froude number (Fr)
R	Ratio of inertial force to compressibility force	3.	Prandtl number (Pr)
S	Ratio of inertial force to gravity force	4.	Mach number (Ma)

****The options are missing and are to be included.**

Solution:

(P) Ratio of inertia force to viscous force is Reynold's number (Re) (1).

(Q)Ratio of momentum diffusivity to thermal efficiency is Prandtl number (Pr) (3).

(R) Ratio of inertia force to compressibility force is Mach number (Ma) (4).

(S) Ratio of inertia force to gravity force is Froude number (Fr) (2).

Hence, the answer is (D).

Q3. In the context of boundary layers, which one of the following statement is FALSE?

(A) It is a frictional layer, close to the body.

(B) It is a region where fluid flow is irrotational

(C) It is region across which the pressure gradient is negligible

(D) It is diffusion layer of vorticity

Solution:

The flow is irrotational in the region outside of the boundary layer, but it is rotational inside the boundary layer.

Hence, the statement (B) is false.

Q4. Consider an ideal fluid flow past a circular cylindrical shown in the figure below. The peripheral velocity at a point P on the surface of the cylinder is-

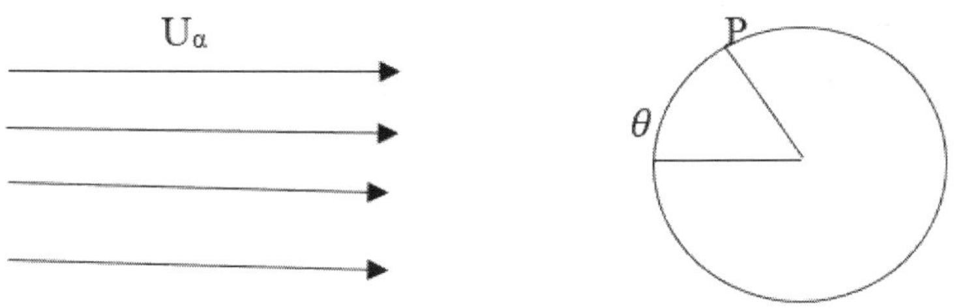

(A)0 (B) U_α (C) $U_\alpha \sin\theta$ (D) $2U_\alpha \sin\theta$

Solution:

The peripheral velocity at a point on the surface of the cylinder is given by, $V = 2U_\infty \sin\theta$

Hence, the answer is (D).

Q5. The rheological diagram depicting the relation between shear stress and shear rate for different types of fluids is shown in fig below-

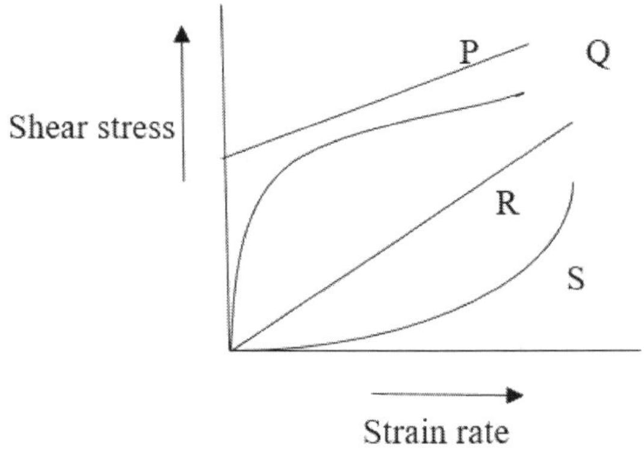

The most suitable relation for flow of tooth paste being squeezed out of the tube is given by the curve-

P (B) Q (C) R

(D) S

Solution:

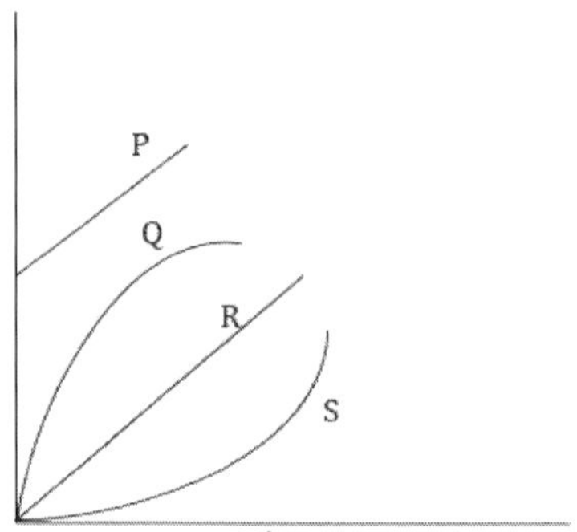

Figure

In the figure

P → Bingham plastic

Q → Pseudo plastic

R → Newtonian

S → Dilatant

Toothpaste is an example of Bingham plastic fluid, which requires a minimum stress before its deformation starts.

Hence, the answer is (A).

Q6. The diverging limb of a venturimeter is kept longer than the converging limb to-

(A) ensure that the flow remains longer.

(B) avoid separation.

(C) ensure that the flow remains turbulent.

(D) avoid formation of boundary layer.

Solution:

The diverging limb of the venter meter is kept longer than the converging limb to avoid separation.

Hence, the answer is (B).

Q7. The length scale of a model is kept as 1:64. The prototype fluid is water. Viscous and gravity forces are equally dominant in prototype. The required kinematic viscosity (m^2/s) of the fluid used in the model is-

(A)0.100E-07 (B) 0.156E-07

(C) 0.195E-08 (D) 0.125E-07

Solution:

Given:-

Length scale of model, $\frac{L_m}{L_p} = \frac{1}{64}$

Since the viscous and gravity forces are equally dominant, hence the Reynold's number and Froude number for both model and prototype are equal.

i.e $\text{Re}_m = \text{Re}_p$ and $\text{Fr}_m = \text{Fr}_p$

Now, $\text{Fr}_m = \text{Fr}_p$

$$\Rightarrow \frac{V_m}{\sqrt{L_m g}} = \frac{V_p}{\sqrt{L_p g}}$$

$$\Rightarrow \frac{V_m}{V_p} = \sqrt{\frac{L_m}{L_p}}$$

$$= \sqrt{\frac{1}{64}}$$

$$= \frac{1}{8}$$

Also, $\text{Re}_m = \text{Re}_p$

$$\Rightarrow \frac{V_m L_m}{\nu_m} = \frac{V_p L_p}{\nu_p}$$

$$\Rightarrow \nu_m = \frac{V_m}{V_p} \times \frac{L_m}{L_p} \times \nu_p$$

$$= \frac{1}{8} \times \frac{1}{64} \times 10^{-6} \text{ m}^2/\text{s} \quad \left(\because \text{the kinematic viscosity of water is } 10^{-6} \text{ m}^2/\text{s}\right)$$

$$= 0.1953 \times 10^{-8} \text{ m}^2/\text{s}$$

Hence, the answer (B) is correct.

Q8. Let ϕ and ψ represents, respectively the velocity potential and stream function of a flow field of an incompressible fluid. Which of the following statements are true?

P: ϕ exists for irrotational flows only

Q: ψ exists for both irrotational and rotational flows

R: ϕ exists for rotational flows only

S: ϕ exists for both rotational and irrotational flows

(A) P, R (B) Q, S (C) Q, R (D) P, Q

Solution:

Velocity potential (ϕ) exists only for an irrotational flow and stream function (Ψ) exists for both a rotational and an irrotational flow.

Hence, the answer is (D).

Q9. Two models M_1 and M_2 have equal volumes and are made of steel. Model M_1 is an aerofoil (a streamlined body) and model M_2 is a sphere (a bluff body). Both models are dropped in two identical short jars filled with honey simultaneously. Which of the following statement is true?

(A) M_1 reaches the bottom earlier than M_2

(B) M_2 reaches the bottom earlier than M_1

(C) Both models reach the bottom at the same time

(D) Both models float on the surface

Solution:

Drag force acting on an aerofoil (streamlined body) is always lower than that on a bluff body.

It is known that drag force acts in the direction opposite to the motion of body. Since model M_1 is aerofoil is a streamlined body and model M_2 is a bluff body, hence the drag force on M_1 is lower than that on M_2. Therefore, M_1 will move faster than M_2 and will reach earlier than M_2.

Hence, the answer is (A).

Two marks questions-

Q10. The wind is blowing east-west at time $t<T$, and switches to south-north at $t = T$. At, $t<T$, which of the following curves represent streaklines?

A)

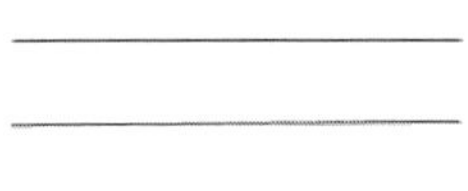

B)

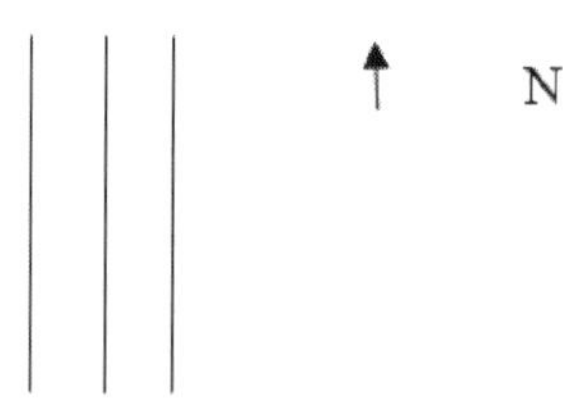

C)

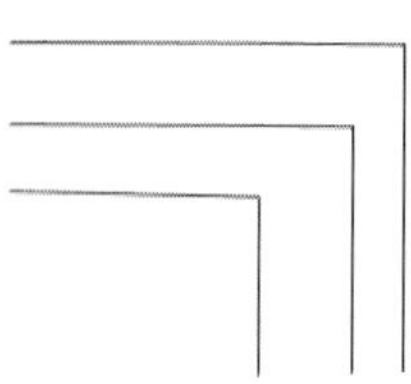

D)

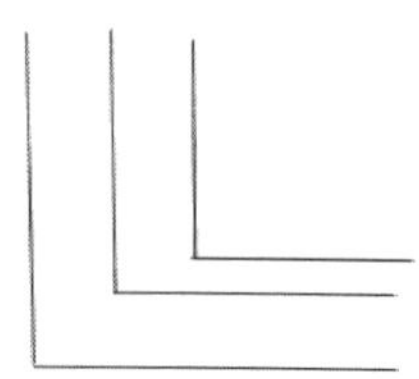

Solution:

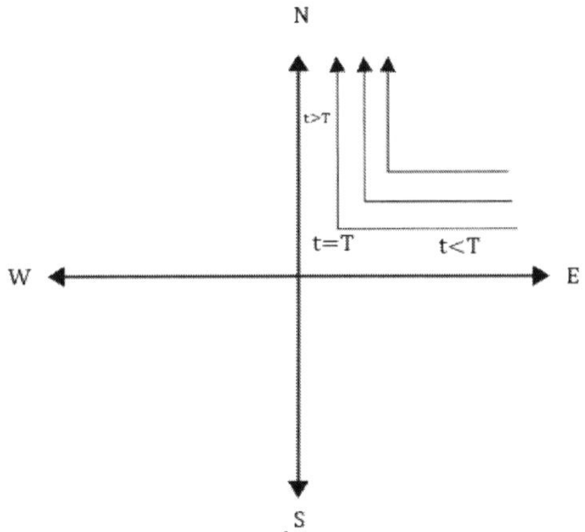

Figure

A streak line at any instant of time is the locus of the temporary locations of all the particles that have crossed a fixed point in the flow field. At $t<T$, the wind is blowing in the East-West direction and is yet to blow in the south-north direction.

Hence, the answer is (A).

Q11. A cricket hall comprises of a seam running along its central section, which is essentially stitching of two hemi-spheres. The seam creates additional roughness. The bowler releases the ball with seam orientation as shown in the figure below-

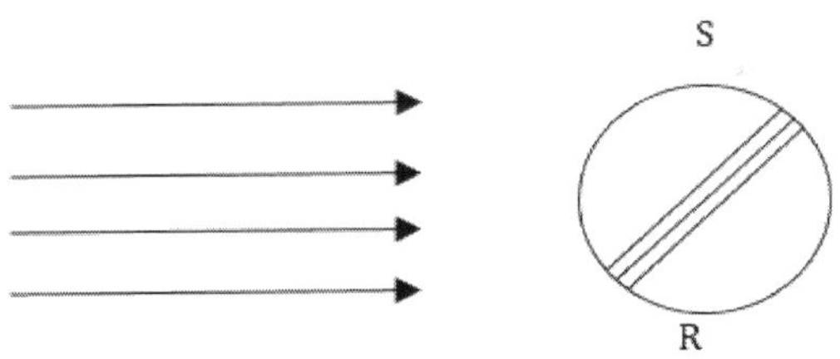

This would result in an out swinger with side forces on the cricket ball. These side forces on the ball are attributed to-

(A) flow having a laminar boundary separation on both sides.

(B) flow having turbulent boundary layer separation on both sides.

(C) flow having a laminar boundary layer separation on the side "S" and a turbulent boundary layer separation on the side "R".

(D) flow having a laminar boundary layer separation on the side "R" and a turbulent boundary layer separation on the side "S".

Solution:

As shown in the figure, the seam faces the incoming fluid (air) in the side "R" earlier than that in the side "S". Therefore, the seam in the side "R" will create a turbulent flow while the flow in the side "S" will still remain laminar before it reaches the seam.

Hence, the answer is (C).

Q12. Ancients have designed water clocks based upon the head of the water in a circular section container with a hole at the bottom shown in the figure. The radius(r) varies as a function of head (H) to maintain a constant rate of decline of H.

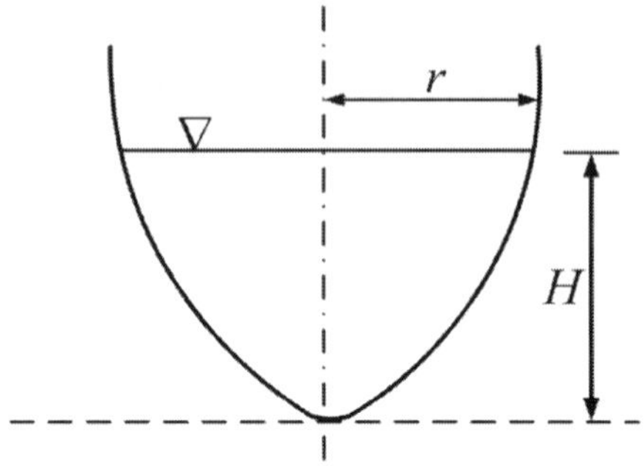

The relation between H and r is

(A) H is proportional to r
(B) H is proportional to r^2
(C) H is proportional to r^3
(D) H is proportional to r^4

Solution:

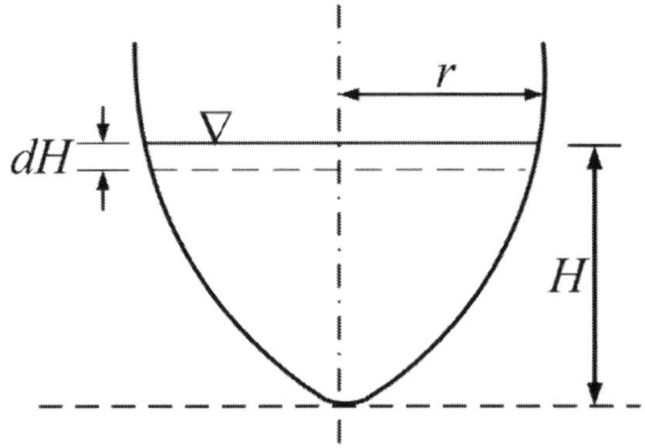

Fig. 1

Let at any instant, the level of water in the container is *H*. Let us assume that the water level decreases by an amount *dH* in a small time interval *dt*, as shown in Fig. 1.

Given that the rate of decline of head *H* is constant, i.e. $\frac{dH}{dt} = \text{constant}$.

Therefore, volumetric flow rate in the container w.r.t. Fig. 1 is given by

$$Q_1 = A_1 \times \frac{dH}{dt}$$

$$\Rightarrow Q_1 = \pi r^2 \frac{dH}{dt} \qquad ...(i)$$

Now, volumetric flow rate at the bottom (hole) is given by

$$Q_2 = A_2 V$$

$$\Rightarrow Q_2 = A_2 \sqrt{2gh} \qquad ...(ii)$$

Further, from the continuity equation we have $Q_1 = Q_2$

$$\pi r^2 \frac{dH}{dt} = A_2 \sqrt{2gH}$$

$$\text{i.e.} \Rightarrow H = \frac{\pi^2}{2gA_2^2} \times \frac{dH}{dt} \times r^4$$

$$\Rightarrow H \propto r^4 \left(\because \frac{dH}{dt} = \text{constant} \right)$$

Hence, the answer is (D).

Q13. A 20 cm diameter pipe carries water discharge of π/100 m³/s. The pipe is bent through an angle of 30⁰ in the horizontal plane as given in the figure below.

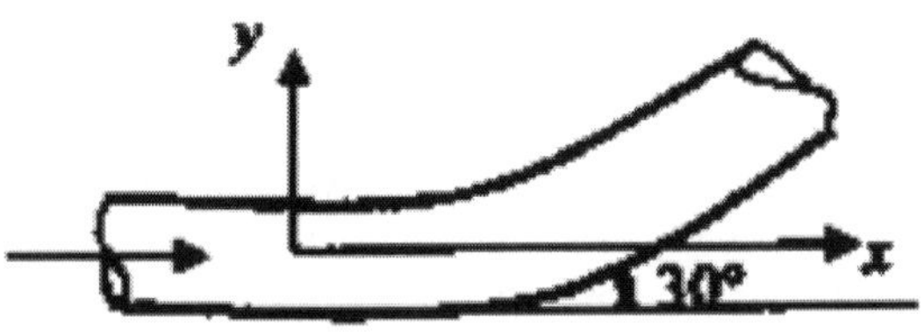

Neglecting friction, the force (N) exerted by water on the bend in x-and y-direction respectively, are-

(A) 4.21 and -15.71 (B) -4.21 and 15.71

(C) 15.71 and -27.2 (D) 4.21 and 15.71

Solution:

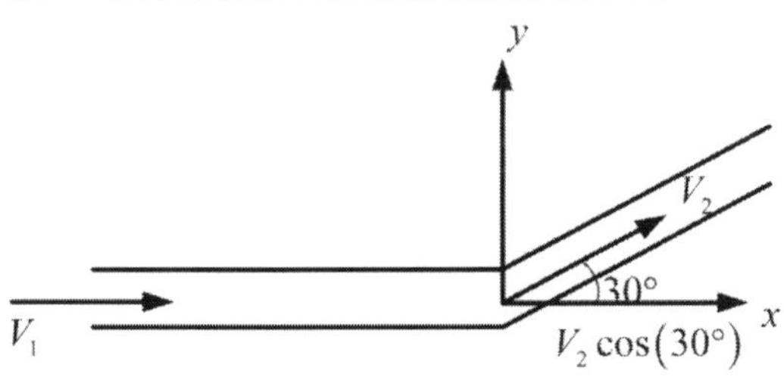

Fig. 1

Given:

Diameter of the pipe, $d = 20$ cm $= 0.20$ m

Rate of discharge, $Q = \dfrac{\pi}{100}\ \text{m}^3/\text{s}$

From continuity equation, we have

$$Q = AV = \frac{\pi}{4}d^2V$$

Hence, $V = \dfrac{4Q}{\pi d^2}$

$$\therefore V_1 = V_2 = \frac{4}{\pi}\times\frac{\pi}{100}\times\frac{1}{0.2\times0.2}\ \text{m/s} = 1\ \text{m/s}$$

Now, x-component of the force exerted by water on the bend is given by

$$F_x = \rho Q\{V_1 - V_2\cos(30°)\}$$

$$= 1000\times\frac{\pi}{100}\{1 - 1\times\cos(30°)\}\ \text{N}$$

$$= 4.21\ \text{N}$$

and y-component of the force exerted by water on the bend is given by

$$F_y = \rho Q\{0 - V_2\sin(30°)\} = 1000\times\frac{\pi}{100}\times\{0 - 1\times\sin(30°)\}\ \text{N} = \text{-}15.7\ \text{N}$$

Hence, the answer is (A).

Q14. A differential U-tube manometer with mercury as the manometric fluid is used to measure the pressure difference between two sections P and Q in a horizontal pipe carrying water at steady state as shown in the figure below. If the difference in mercury levels in the two limbs of the manometer is 0.75 m, in pressure (kPa) between sections P and Q is-

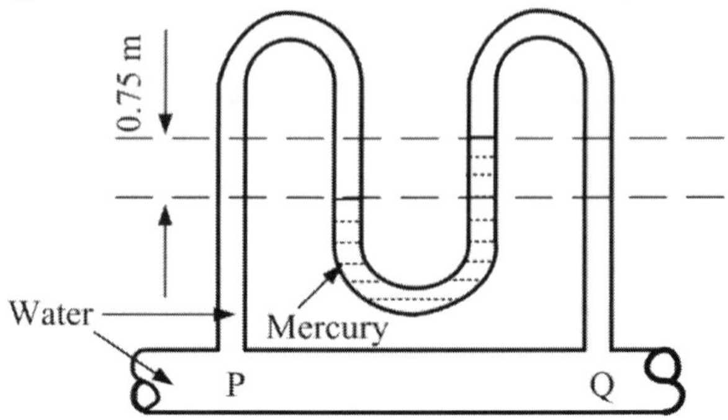

(A) 49.275 (B) 94.275 (C) 9.4275 (D) 492.75

Solution:

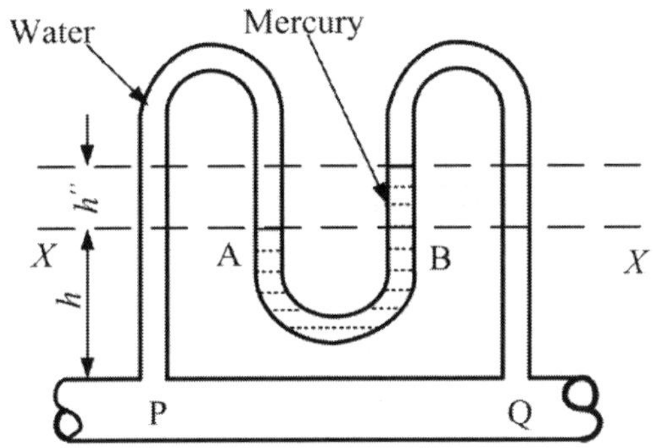

Fig. 1

Equating the pressures at the two points A and B on the reference line *X-X*, we have the monomeric equation

$$p_P - \rho_w gh = p_Q - \rho_W gh - \rho_W gh'' + \rho_{Hg} gh''$$

$$\Rightarrow p_P - p_Q = \left(\rho_{Hg} - \rho_W\right) gh''$$

$$= (13600 - 1000) \times 10 \times 0.75 \text{ Pa}$$

$$= 94500 \text{ Pa}$$

$$= 94.5 \text{ kPa}$$

Hence, the answer is (B).

Q16. A 20cm cubical box slides on oil (mass density = 800kg/), over a large plane surface with a steady state velocity of 0.4 m/s. The plane surface is inclined at an angle of 30 degree with the horizontal plane. The oil film between the block and the plane surface is 0.4mm thick. The weight of the cubical box is 64N. The kinematic viscosity of the oil is:

(A) 0.8 Pa.s (B) 0.001 m^2/s (C) 1.6 Pa.s (D) 0.002 m^2/s

Solution:

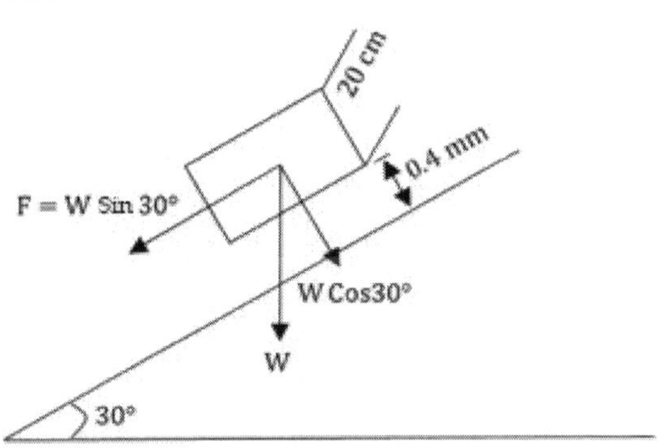

Fig. 1

Given:-

Dimension of box, $a = 20$ cm $= 0.20$ m

Density of oil, $\rho =$ kg/m^3

Velocity of the block, $V = 0.4$ m/s

Angle of inclination, $\theta = 30°$

Film, thickness, $d = 0.4$ mm $= 0.4\times10^{-3}$ m

Weight of the box, $W = 64$ N

Now, from Newton's law of viscosity

$$\frac{F}{A} = \mu\frac{du}{dy}$$

$$\Rightarrow \frac{W\sin(30°)}{a^2} = \mu\frac{V}{d}$$

$$\Rightarrow \frac{64\sin(30°)}{0.2^2} = \mu\times\frac{0.4}{0.4\times10^{-3}}$$

$$\Rightarrow \mu = 0.8 \text{ Pa.s}$$

Therefore, the kinematic viscosity,

$$\nu = \frac{\mu}{\rho} = \frac{0.8}{800}\text{ m}^2/\text{s} = 0.001\text{ m}^2/\text{s}$$

Hence, the answer is (B)

Common data for questions for question 17 & 18

A 60% efficient pump is installed in pipe of diameter 20 cm to lift water from a sump to overhead tank at discharge rate of π/100 /s. Free surface level in the overhead tank is 20 m higher than the free surface level in the sump. The inclusive head loss (not including the lift) in the suction and delivery sides of the pump are 2 times and 28 times the velocity head respectively.

Q17. The power W supplied to the pump is

(A) 10476.2 (B) 6285.7 (C) 6757.1 (D) 11231.9

Solution:

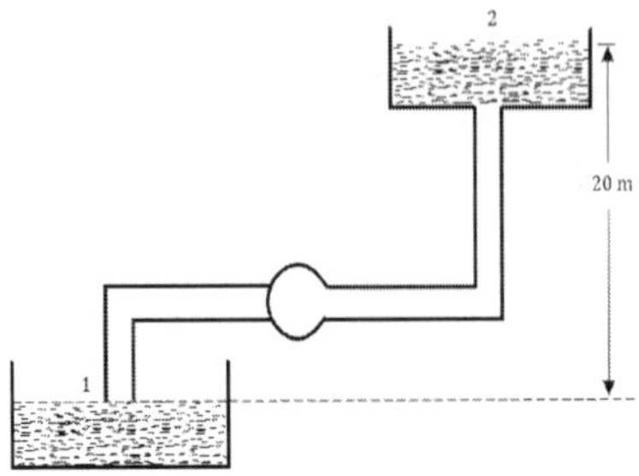

Fig. 1

Given:-

Efficiency of pump, $\eta = 60\%$

Diameter of the pump, $d = 20$ cm = 0.20 m

Discharge rate, $Q = \dfrac{\pi}{100}\ \mathrm{m^3/s}$

From continuity equation, we have the velocity in the pipe

$$V = \frac{4Q}{\pi d^2} = \frac{4}{\pi \times 0.2^2} \times \frac{\pi}{100}\ \mathrm{m/s} = 1\ \mathrm{m/s}$$

Now, applying the principle of energy conservation between point 1 and point 2, we have.

$$\frac{p_1}{\rho g} + \frac{V_1^2}{2g} + z_1 + H = \frac{p_2}{\rho g} + \frac{V_2^2}{2g} + z_2 + h_l \qquad \text{(i)}$$

where, $p_1 = p_2 = p_{atm}, z_2 - z_1 = 20\ \mathrm{m}, V_1 = V_2 = V$

$$h_l = \text{Total head loss} = \frac{2V^2}{2g} + \frac{28V^2}{2g} = \frac{30V^2}{2g} = \frac{30 \times 1^2}{2 \times 10}\ \mathrm{m} = 1.5\ \mathrm{m}$$

and H is the energy head transferred to the water by the pump.

Hence, from eq. (i)

$$\begin{aligned} H &= \frac{p_2 - p_1}{\rho g} + \frac{V_2^2 - V_1^2}{2g} + (z_2 - z_1) + h_l \\ &= (0 + 0 + 20 + 1.5)\ \mathrm{m} \\ &= 21.5\ \mathrm{m} \end{aligned}$$

Therefore, the power delivered by the pump is given by

$$\begin{aligned} P_0 &= \rho g Q H \\ &= 1000 \times 10 \times \frac{\pi}{100} \times 21.53\ \mathrm{W} \\ &= 6751\ \mathrm{W} \end{aligned}$$

If P_i is the input power required by the pump, then the pump efficiency is given by

$$\eta = \frac{P_0}{P_i} \times 100$$

$$\Rightarrow 60 = \frac{6751}{P_i} \times 100$$

$$\Rightarrow P_i = 6751 \times \frac{100}{60}\text{ W}$$

$$= 11251.6\text{ W}$$

Hence, the answer is (D).

Q18. The suction side of the pump is located L m above the free surface level in the pump. The minimum permissible pressure in the pipeline of the suction side of the pump is 8 m of water below atmospheric pressure. The maximum possible value of L is

(A) 20.00

(B) 8.00

(C) 7.85

(D) 5.00

Solution:

Applying the principle of energy conservation between point 1(free surface level) and point 2 (the suction point of the pump), we have

$$\frac{p_1}{\rho g} + \frac{V_1^2}{2g} + z_1 = \frac{p_2}{\rho g} + \frac{V_2^2}{2g} + z_2 + (h_l)_{\text{in}} \qquad \ldots\text{(i)}$$

where, $\frac{p_1 - p_2}{\rho g} \leq 8\text{ m}, z_2 - z_1 = L\text{ m}, V_1 = V_2 = V$

and $(h_l)_{\text{in}}$ = Head loss at the inlet $= \frac{2V^2}{2g} = \frac{2V^2}{2g} = \frac{2\times 1^2}{2\times 10}\text{ m} = 0.1\text{ m}$

Hence, from eq. (i)

$$z_2 - z_1 = \frac{p_1 - p_2}{\rho g} + \frac{V_1^2 - V_2^2}{2g} - (h_l)_{\text{in}}$$

$$\Rightarrow L \leq 8 + 0 - 0.1$$

$$\Rightarrow L \leq 7.9$$

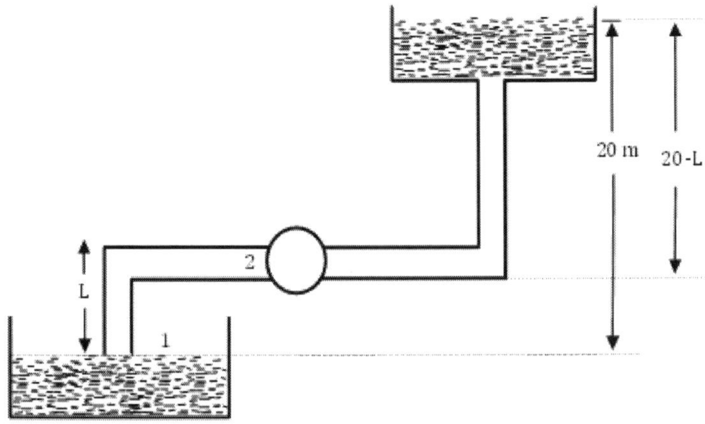

Fig. 1

Hence, the answer is (C).

Common data for question 19 & 20

The velocity for two- dimensional fluid flow is as follows:

$$U = U_0 \frac{x}{L} V = -U_0 \frac{y}{L}$$

where, U_o and L are, respectively, the characteristic velocity and length

Q19. If the L= 0.2 m and the resultant of total acceleration in X and Y direction at (X=L, Y =L) is 10 m/, the magnitude of (m/s) is

1.414 (B) 2.38 (C) 1.19 (D) 11.90

Solution:

Given:

$$U = U_0 \frac{x}{L} \text{ and } V = -U_0 \frac{y}{L}$$

Now, *x*-component of acceleration is given by,

$$a_x = \frac{\partial U}{\partial t} + U\frac{\partial U}{\partial x} + V\frac{\partial U}{\partial y}$$

$$= 0 + U_0 \frac{x}{L} \times \frac{U_0}{L} + \left(-U_0 \frac{y}{L}\right) \times 0$$

$$= \frac{U_0^2}{L^2} x$$

At, $x = L$ and $y = L$, we have

$$a_x = \frac{U_0^2}{L^2}L = \frac{U_0^2}{L}$$

And y-component of acceleration is given by

$$a_y = \frac{\partial V}{\partial t} + U\frac{\partial V}{\partial x} + V\frac{\partial V}{\partial y}$$

$$= 0 + U_0\frac{x}{L}\times 0 + \left(-U_0\frac{y}{L}\right)\times\left(-\frac{U_0}{L}\right)$$

$$= \frac{U_0^2}{L^2}y$$

At, $x = L$ and $y = L$, we have

$$a_y = \frac{U_0^2}{L^2}L = \frac{U_0^2}{L}$$

∴ Magnitude of the total acceleration is given by

$$a = \sqrt{a_x^2 + a_y^2}$$

$$= \sqrt{\left(\frac{U_0^2}{L}\right)^2 + \left(\frac{U_0^2}{L}\right)^2}$$

$$= \frac{U_0^2}{L}\sqrt{2}$$

At $L = 0.2$ m, $a = 10$ m/s^2

$$\therefore \frac{U_0^2}{0.2}\times\sqrt{2} = 10$$

$$\Rightarrow U_0 = 1.19\ \text{m/s}$$

Hence, the answer is (C).

Q20. The above fluid flow can be described as

(A) rotational and compressible (B) irrotational and compressible

(C) rotational and incompressible (D) irrotational and incompressible

Solution:

The continuity equation for a two-dimensional (2D) incompressible flow is

$$\frac{\partial u}{\partial x} + \frac{\partial v}{\partial y} = 0$$

For the present case

$$\frac{\partial u}{\partial x}+\frac{\partial v}{\partial y}=\frac{\partial}{\partial x}\left(U_0\frac{x}{L}\right)+\frac{\partial}{\partial y}\left(-U_0\frac{y}{L}\right)$$

$$=\frac{U_0}{L}-\frac{U_0}{L}$$

$$=0$$

Hence the velocity profile satisfies the continuity equation for a 2D incompressible flow.

Also, the angular velocity is

$$\omega_z=\frac{1}{2}\left(\frac{\partial v}{\partial x}-\frac{\partial u}{\partial y}\right)$$

$$=\frac{1}{2}\left[\frac{\partial}{\partial x}\left\{-U_0\frac{y}{L}\right\}-\frac{\partial}{\partial y}\left\{U_0\frac{x}{L}\right\}\right]$$

$$=0$$

Hence the flow is irrotational.

Hence, the answer is (D).

Statement for linked answer question 21 and 22

The boundary layer formation over a flat plate is shown in the figure below , The variation of horizontal velocity (u) Y at any X along the plate in the boundary layer is approximated as: u=Psin(Qy)+ R

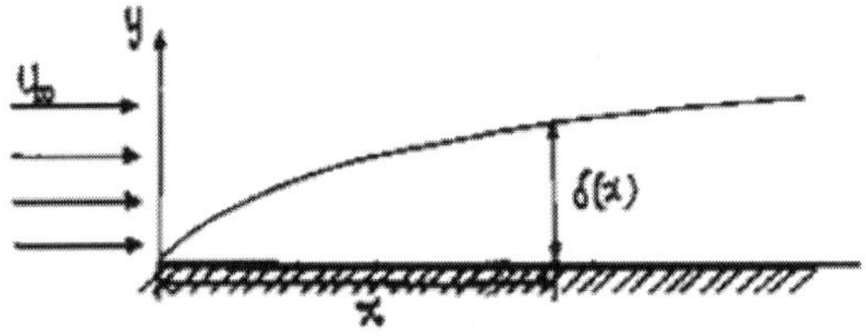

Q21. The most acceptable boundary conditions are

(A) at $y=0,\ u=0;$ at $y=\delta,\ u=U_\infty;$ at $y=0, \frac{du}{dy}=0$

(B) at $y=0,\ u=U_\infty;$ at $y=\delta,\ u=U_\infty;$ at $y=0, \frac{du}{dy}=0$

(C) at $y=0,\ u=0;$ at $y=\delta,\ u=U_\infty;$ at $y=\delta, \frac{du}{dy}=0$

(D) at $y=0,\ u=U_\infty;$ at $y=\delta,\ u=U_\infty;$ at $y=\delta, \frac{du}{dy}=0$

Solution:

Variation of velocity is given as $u = P\sin(Qy) + R$

The most acceptable boundary conditions for a boundary layer is

$$\text{at } y = 0, u = 0;\ y = \delta, u = U_\infty;\ y = \delta, \frac{du}{dy} = 0$$

Hence, the answer is (C).

Q22. The expression for P, Q, and R are

(A) $P = 0;\ Q = \frac{\pi}{2\delta};\ R = U_\infty$

(B) $P = U_\infty;\ Q = \frac{\pi}{2\delta};\ R = 0$

(C) $P = U_\infty;\ Q = \frac{\pi}{2\delta};\ R = 0$

(D) $P = U_\infty;\ Q = \frac{\pi}{2\delta};\ R = 0$

Solution:

We have $u = P\sin(Qy) + R$

Now, at $y = 0, u = 0,$ i.e., $0 = P\sin(Q\times 0) + R \Rightarrow R = 0$

$$\Rightarrow U_\infty = P\sin(Q\delta)\ \ (\because R = 0) \qquad \text{...(i)}$$

Also, at $y = \delta, \frac{du}{dy} = 0,$

i.e.,

$$PQ\cos(Q\delta) = 0$$

$$\Rightarrow \cos(Q\delta) = 0$$

$$\Rightarrow Q\delta = \frac{\pi}{2}$$

$$\therefore Q = \frac{\pi}{2\delta} \qquad \text{...(ii)}$$

Now from eq. (i), $U_\infty = P\sin\left(\frac{\pi}{2\delta}\times\delta\right) \Rightarrow P = U_\infty$

Therefore, $P = U_\infty, Q = \frac{\pi}{2\delta}, R = 0$

Hence, the answer is (D).

GATE 2009
FLUID MECHANICS

Q1. Under what conditions is the equation $\Delta.p\vec{V}=0$ valid?

P: Steady incompressible flow

Q: Unsteady incompressible flow

R: Steady compressible flow

S: Unsteady compressible flow

(A) P.Q.R (B) Q.R.S

(C) P.R.S (D) P.Q.S

Solution:

General continuity equation is given by, $\nabla\rho\vec{v}+\dfrac{\partial\rho}{\partial t}=0\rightarrow$ ...(i)

Equation (i) is valid for steady, unsteady, compressible and incompressible flow.

But for unsteady flow, ρ becomes independent of time and $\dfrac{\partial\rho}{\partial t}$ becomes zero.

Therefore, equation (i) reduces to $\nabla\rho\vec{v}=0\rightarrow$ ...(ii)

Thus, equation (ii) is valid for steady, compressible and incompressible flow.

Moreover, for incompressible flow, ρ is constant therefore, equation (i) becomes $\nabla\rho\vec{v}=0$

i.e. irrespective of steady or unsteady, if flow is incompressible, ρ becomes constant and the $\dfrac{\partial\rho}{\partial t}$ becomes zero. Equation (i) reduces to $\nabla\rho\vec{v}=0\rightarrow$ ...(iii)

Therefore, equation (iii) is valid for steady incompressible, unsteady incompressible and steady compressible flow.

Hence, answer is (A).

Q2. Stream function CANNOT be defined for

(A) two dimensional incompressible flow

(B) two dimensional compressible flow

(C) three dimensional incompressible flow

(D) axisymmetric incompressible flow

Solution:

Stream function (Ψ) can be defined only for incompressible flow.

Hence, answer is (B).

Q3. Which one of the following is an irrotational flow?

(A) Free vortex flow
(B) Forced vortex flow
(C) Couette flow
(D) Wake How

Solution:

Free vortex flow is an irrotational flow.

Hence, answer is (A).

Q4. Under strong wind conditions, electrical cables can be subjected to wind-induced oscillations. Which one of the following non-dimensional numbers is relevant to this problem?

(A) Froude number
(B) Weber number
(C) Faraday number
(D) Strouhal number

Solution: D

The Strouhal Number is a dimensionless value useful for analyzing oscillating unsteady fluid flow dynamics problems.

The Strouhal Number can be expressed as

$$St = \frac{\omega l}{v}$$

where

St = Strouhal Number

ω = *oscillation frequency*

l = characteristic length

v = flow velocity

The Strouhal Number can be important when analyzing unsteady, oscillating flow problems. The Strouhal Number represents a measure of the ratio of the *inertial forces* due to the *unsteadiness of the flow* or local acceleration to the inertial forces due to changes in velocity from one point to an other in the flow field.

The vortices observed behind a stone in a river, or measured behind the obstruction in a vortex flow meter, illustrates these principles.

Q5. Dimples are made on golf balls for which of the following reasons?

P: to make the ball travel a longer distance

Q: to make the flow over the ball turbulent

R: to make the flow over the ball laminar

S: to create a separated boundary layer flow over the ball

(A) P, Q
(B) Q, S
(C) R, S
(D) P, R

Solution:

Dimples are made on golf balls so that the flow of fluid (air) over the ball becomes turbulent. For turbulent flow, point of separation of boundary layer shifts towards downward and small wake region is formed. Thus, drag force on the ball decreases and ball moves faster and moves longer distance.

Hence, answer is (A).

Q6. In a 2-D boundary layer flow, x and y are the streamwise and wall-normal coordinates, respectively. If denotes the velocity along the x direction, which one of the following represents the condition at the point of flow separation?

(A) $\frac{\partial u}{\partial x}=0$ (B) $\frac{\partial u}{\partial y}=0$ (B) $\frac{\partial^2 u}{\partial x^2}=0$ (C) $\frac{\partial^2 u}{\partial y^2}=0$

Solution:

At the point of boundary layer separation, following are the conditions:-

$$\frac{\partial u}{\partial y}=0$$

$$\frac{\partial^2 u}{\partial y^2}>0$$

$$\frac{\partial p}{\partial x}>0$$

Shear stress = 0

Hence, answer is (B).

Q7. Which one among the following boundary layer flows is the LEAST susceptible to flow separation?

(A) turbulent boundary layer in a favourable pressure gradient

(B) laminar boundary layer in a favourable pressure gradient

(C) turbulent boundary layer in an adverse pressure gradient

(D) laminar boundary layer in an adverse pressure gradient

Solution:

Boundary layer separation occurs due to the loss of kinetic energy possessed by the flow. Due to loss of K.E, flow does not drag on the surface through larger distance and gets separated.

Turbulent flow has more kinetic energy and it can drag through larger distance on the surface therefore, turbulent flow is susceptible to separation.

Moreover, negative pressure gradient or favourable pressure gradient favours the flow and adverse pressure gradient opposes the flow, which leads to separation.

Thus, turbulent boundary layer in a favourable pressure gradient is least susceptible to flow separation.

Hence, answer is (A).

Q8. Air from the blower of a hairdryer flows between two identical elliptical cylinders suspended freely, for two cases shown in the figure. The cylinders would move

(A) away from each other for Case 1 and towards each other for Case 2

(B) towards each other for Case 1 and away from each other for Case 2

(C) away from each other for Case 1 and away from each other for Case 2

(D) towards each other for Case 1 and towards each other for Case 2

Solution:

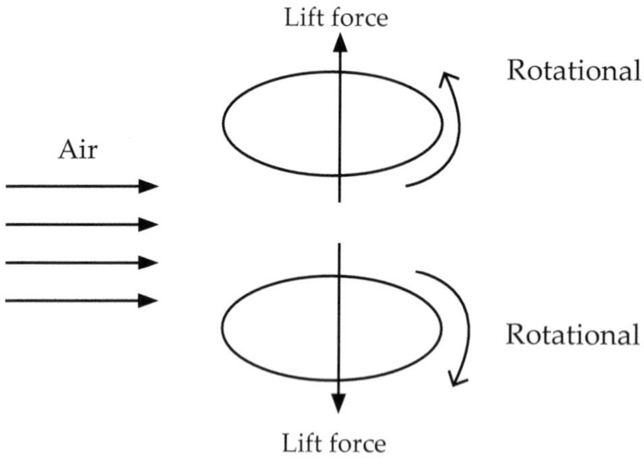

When air flows between two cylinders, upper cylinder will rotate anticlockwise, and lower cylinder will rotate clockwise. Therefore, according to Magnus effect, lift force will act in upward direction on upper cylinder and in downward direction on lower cylinder. Therefore, cylinder will move away from each other in both cases.

Hence, answer is (C).

Q. 9 to Q.24 carry two marks each.

Q9. A 40 cm cubical block slides on oil (viscosity = 0.80 Pa.s), over a large plane horizontal surface. If the oil film between the block and the surface has a uniform thickness of 0.4 mm, what will be the force required to drag the block at 4 m/s? Ignore the end effects and treat the flow as two dimensional.

(A) 1280 N (B) 1640 N (C) 1920 N (D) 2560 N

Solution:

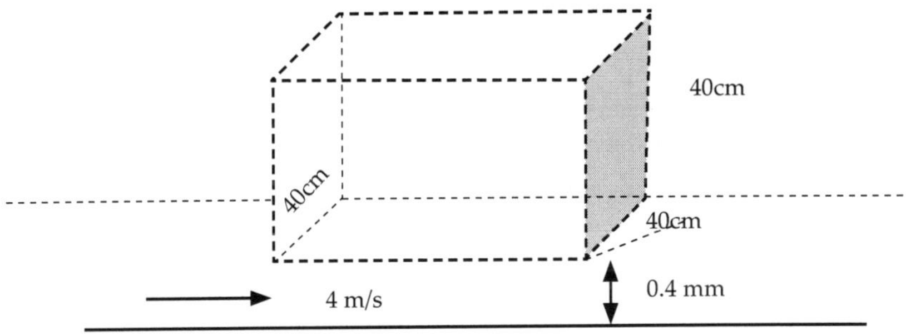

Given:

Dimension of the block, a = 40 cm

Viscosity of the oil, $\mu = 0.80$ Pa. s

Thickness of the film. D = 0.4 mm = 0.4×10^{-3} m

Velocity of the block, v = 4 m/s

From Newton's law of viscosity, we have

$$\frac{F}{A} = \mu\frac{du}{dy}$$

$$F = \mu A\frac{du}{dy}$$

$$F = 0.80\times0.4\times0.4\times\frac{4}{0.4\times10^{-3}}$$

$$F = 1280\text{N}$$

Hence, answer is (A).

Q10. For a floating body, G, B, and M represent the centre of gravity, centre of buoyancy, and the metacentre, respectively. The body will be stable if

(A) G is located above B (B) B is located above M

(C) M is located above B (D) M is located above G

Solution:

For a floating body, to be in stable equilibrium, metacenter M should be located above center of gravity G.

Hence, answer is (D).

Q11. A nozzle has inlet and outlet diameters of 10 cm and 5 cm, respectively. If it discharges air at a steady rate of 0.1 m³/s into the atmosphere, the gauge pressure (static) at the nozzle inlet will be

(A) 1.26 kPa (B) 1.46 kPa (C) 3.52 kPa (D) 3.92 kPa

Solution:

Given:

Inlet diameter, d_1 = 10 Cm, = 0.1 m

Outlet diameter, d_2 = 5 Cm = 0.05 m

Flow rate of air, Q = 0.1 m^3/s

Pressure of the outlet, P_2 = 0 atm (gauge) = 0 kPa

Now, from continuity equation we have

$$Q = A_1V_1 = A_2V_2$$

$$Q = \frac{\pi d_1^2}{4} v_1 = \frac{\pi d_2^2}{4} v_2$$

$$v_1 = \frac{4Q}{\pi d_1^2} \text{ and } v_2 = \frac{4Q}{\pi d_2^2}$$

$$\therefore v_1 = \frac{4\times 0.1}{3.14\times(0.1)^2} = 12.74 \text{ m/s and } v_2 = \frac{4\times 0.1}{3.14\times 0.05\times 0.05} = 50.95 \text{ m/s}$$

Now, from the Bernoulli's equation, we have

$$\frac{P_1}{\rho g} + \frac{v_1^2}{2g} + Z_1 = \frac{P_2}{\rho g} + \frac{V_2^2}{2g} + Z_2$$

$$0 + \frac{(12.74)^2}{2\times 9.81} = \frac{P_2}{\rho g} + \frac{(50.95)}{2\times 9.81}$$

$$8.27 = \frac{P_2}{1.2\times 9.81} + 132.31$$

$$-124.04 = \frac{P_2}{1.2\times 9.81}$$

$$P_2 = -1460.198 \text{ Pa} = -1.46 \text{ kPa}$$

Hence, answer is (B).

Q12. Consider incompressible flow through a two-dimensional open channel. At a certain section A-A, the velocity profile is parabolic. Neglecting air resistance at the free surface, find the volume flow rate per unit width of the channel.

(A) 10 m²/s (B) 13.33 m²/s (C) 20 m²/s (D) 33.33 m²/s

Solution:

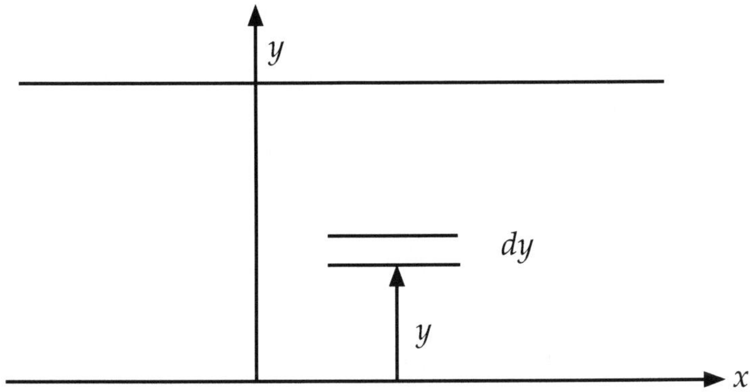

Let the velocity profile be given by,

$u = ay^2 + by + c$

Where, a, b and c are the constant.

Now, from boundary conditions we have

At y = 0, u = 0 and at y = 2m, u = 10 m/s

$\therefore$ 0 = a×0 + b×0 + c

c = o

And, 10 = a×2× + b×2

10 = 4a + 2b → (i)

Now, we have $\frac{du}{dy} = 2ay + b$

At $y = 2\text{m}, \frac{du}{dy} = 0$

i.e 2a × 2 + b = 0

4a + b = 0 → ...(ii)

Solving (i) and (ii) we get, b = 10 and $a = -5/2$ Velocity profile is given by,

$u = \frac{-5}{2}y^2 + 10y$

Now, volumetric flow rate through small strip of the thickness dy is given by,

dQ = u.1.dy (Assuming width unity)

Hence, answer is (B).

$$dQ=\left(\frac{-5}{2}y^2+10y\right)dy$$

$$Q=\frac{-5}{2}\int_0^2 y^2dy+10\int_0^2 y^2dy$$

$$Q=\frac{-5}{2}\frac{\left[y^3\right]_0^2}{3}+10\frac{\left[y^2\right]_0^2}{2}$$

$$Q=\frac{-5}{2\times3}\times2\times2\times2+\frac{10}{2}\times2\times2$$

$$Q=\frac{-20}{3}+20$$

$$Q=13.33\,m^3/s$$

Q13. Water flows from an open vertical cylindrical tank of 20 cm diameter through a hole of 10cm diameter. What will be the velocity of water flowing out of the hole at the instant when the water level in the tank is 50 cm above the hole? Ignore unsteady effects.

(A) 3.16 m/s (B) 3.26 m/s (C) 3.36 m/s (D) 3.46 m/s

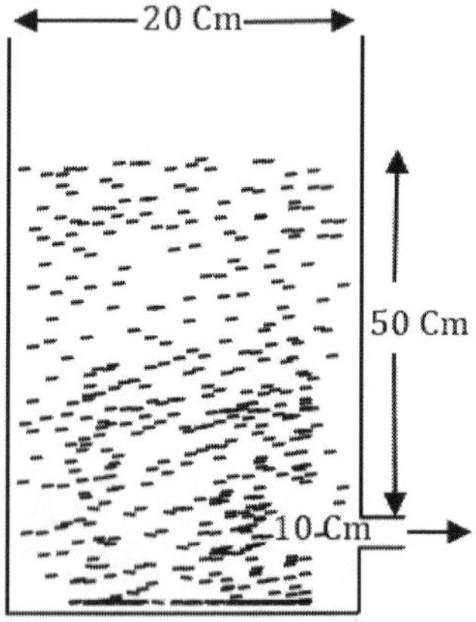

Solution:

Given:

Diameter of the tank, d = 20 Cm = 0.20 m

Diameter of the hole, d_0 = 10 Cm = 0.10 m

Water level in the tank, h = 50 Cm = 0.50 m

Now, velocity of water through the hole is given by,

$$v=\sqrt{2gh}=\sqrt{2\times9.81\times0.50}=3.125 \text{ m/s}$$

Hence, answer is (A).

Q14. In the manometer shown in the figure, the pressure p_A of the gas inside bulb A is approximately,

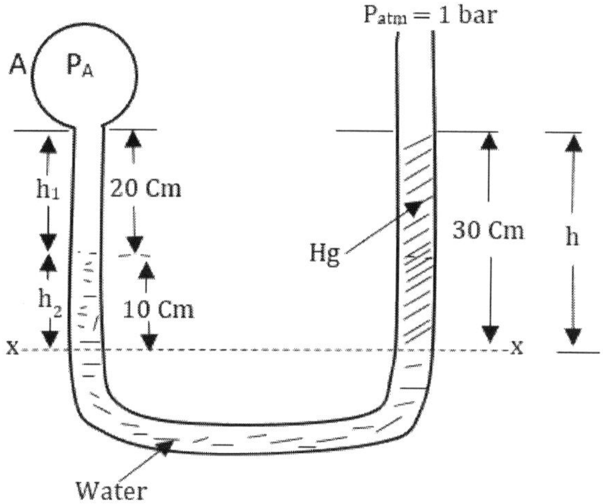

(A) 0.8 bar (B) 1.2 bar (C) 1.4 bar (D) 1.6 bar

Solution:

Taking X-R as reference line, we have manometer equation,

$$P_A + \rho_a g h_1 + \rho_w g h_2 = \rho_{Hg} g h + P_{atm}$$

$$P_A + 1.2 \times 9.81 \times 0.20 + 1000 \times 9.81 \times 0.10 = 13600 \times 9.81 \times 0.30 + 1 \times 1.01 \times 10^5$$

$$P_A + 983.35 = 141349.8$$

$$P_A = 1.4035 \times 10^5 \text{ pa}$$

$$P_A = 1.4036 \text{ bar}$$

Hence, answer is (C).

Q15. Consider a fully developed laminar flow in a circular pipe. If the diameter of the pipe is halved while the flow rate and length of the pipe are kept constant, the head loss increases by a factor of

(A) 4 (B) 8 (C) 16 (D) 32

Solution: D

The loss of head resulting from the flow of a fluid through a pipeline is expressed by the Darcy Formula

$$h_l = \text{f} \frac{L}{d} \frac{V^2}{2g}$$

where hf is the loss of head (units of length) and the average velocity is V. The friction factor, f, varies with Reynolds number and a roughness factor.

$Q = AV$

$$V = \frac{Q}{A} = \frac{4Q}{\pi d^2}$$

So $h_f \ \alpha \frac{1}{d^5}$

Hence, answer is (D).

Q16. A 1:20 model of a submarine is to be tested in a towing tank containing sea water. If the submarine velocity is 6 m/s, at what velocity should the model be towed for dynamic similarity?

(A) 60 m/s (B) 120 m/s (C) 180 m/s (D) 240 m/s

Solution:

Given:

Length scale of the model, $\frac{l_m}{l_p} = \frac{1}{20}$

Velocity of submarine, $V_p = 6$m/s

Now, for the model and prototype to be in dynamic equilibrium,

Reynold's number for model and prototype must be same.

$$\frac{\rho_m l_{mvm}}{\mu_m} = \frac{\rho_p l_{pvp}}{\mu_p}$$

$$\frac{l_m}{l_p} = \frac{v_p}{v_m} \left(\therefore \rho_m = \rho_p \text{ and } \mu_m = \mu_p\right)$$

$$v_m = \frac{l_p}{l_m} \times v_p = 20 \times 6 = 120 \text{ m/s}$$

Hence, answer is (B).

Q17. An oil droplet (density = 800 kg/m³) is rising in still water at a constant velocity of 1 mm/s. Its radius is approximately

(A) 21 micron

(B) 24 micron

(C) 34 micron

(D) 47 micron

Solution:

Given:

Density of oil, ρ_{oil} = 800 kg/m^3

Velocity of oil droplets, v = 1 mm/s = 0.001 m/s

Now, terminal velocity of droplet of diameter d_p moving with velocity v in water of density ρ_w and viscosity μ is given by

$$v = \frac{(\rho_1 - \rho_w) g d_p^2}{18\mu}$$

$$v = \frac{2}{9}\frac{(\rho_w - \rho_{oil}) g r_p^2}{\mu}$$

$$0.001 = \frac{2}{9}\frac{(1000 - 800) \times 9.81 \times r_p^2}{0.001}$$

$$r_p^2 = 2.2937 \times 10^{-9}$$

$$r_p = 4.789 \times 10^{-15}$$

$$r_p = 4.789 \times 10^{-6}\,\text{m} = 47.89 \text{ micron}$$

Hence, answer is (D).

Q18. Determine the correctness or otherwise of the following Assertion [a] and the Reason [r]:

Assertion [a]: The coefficient of discharge of orifice flow meter is less than that of venturi meter.

Reason [r] : Orifice flow meter is a differential pressure device.

(A) Both [a| and [r] are true and [r] is the correct reason for [a].

(B) Both [a] and [r] are true but [r] is not the correct reason for [a].

(C) Both [a] and [r] are false.

(D) [a] is true but [r] is false.

Solution:

The coefficient of discharge of orifice flow meter is less than that of venturi meter, because due to separation of water in orifice meter, head loss is high while in venturi meter there is a discharging section, which minimizes the separation of water, and hence head loss is minimum and high coefficient of discharge.

Orifice flow meter is a differential pressure device because its principle is based on the pressure difference between free stream in pipe and the vena contracta.

Hence, answer is (B).

Common Data Questions

Common Data for Questions 19 and 20:

A long cylindrical object submerged in still water is moving at a constant speed of 5 m/s perpendicular to its axis, as shown in the figure. Neglect viscous effects and assume free stream pressure to be 100 kPa.

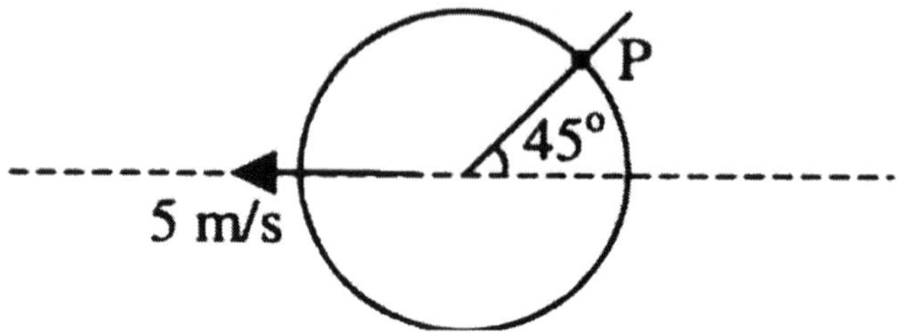

Q19. The fluid velocity at point P with respect to the cylinder will be approximately

(A) 3.5 m/s (B) 5 m/s (C) 7 m/s (D) 10 m/s

Solution:

Given:

Speed of the cylinder, $v = 5$ m/s

Free stream pressure, $P_a = 100$ kPa

Now, fluid velocity at a point P is given by,

$v = 2\gamma \sin\emptyset$

$v = 2\times5\times\sin45 = 7.07$ m/s

Hence, answer is (C).

Q20. The absolute pressure at point P will be approximately

(A) 137 kPa (B) 112 kPa (C) 87 kPa (D) 62 kPa

Solution:

Pressure distribution for flow over a cylinder is given by

$$\frac{p - p_\infty}{\frac{1}{2}\rho v^2} = 1 - 4\sin^2\emptyset$$

Where,

p = pressure on the surface of the cylinder at an angle Ø from centre,

p_∞ = free stream pressure

ρ = density of water

V = free stream velocity

$p - p_\infty = (1-4\sin^2 45) \times 0.5 \times 1000 \times 5 \times 5 = -12500$ Pa = -12.50 kPa

p = -12.500+100 kPa

p = 87.5 kPa

Hence, answer is (C).

Common Data for Questions 21 and 22:

The velocity field fora two dimensional flow is given by: $\vec{V}(x,y,t) = \frac{x}{t}i - \frac{y}{t}j$

Q21. The total acceleration is

(A) $\frac{x}{t}i - \frac{y}{t}j$ (B) $-\frac{x}{t}i - \frac{y}{t}j$

(C) $\frac{2x}{t^2}i$ (D) $\frac{2x}{t^2}j$

Solution:

Given:

Velocity profile, $\vec{v}(x,y,t) = \frac{x}{t}i - \frac{y}{t}j$

$u = \frac{x}{t}$ and $v = \frac{-y}{t}$

Now, x-component of acceleration is given by

$$a_x = u\frac{du}{dx} + v\frac{du}{dy} + \frac{du}{dt}$$

$$a_x = \frac{x}{t} \times \frac{1}{t} + \left(\frac{-y}{t}\right) \times 0 + \frac{t \times 0 - x}{t^2}$$

$$a_x = \frac{x}{t^2} - \frac{x}{t^2} = 0$$

$$a_y = u\frac{du}{dx} + v\frac{dv}{dy} + \frac{dv}{dt}$$

$$a_y = \frac{x}{t} \times 0 + \left(\frac{-y}{t}\right) \times \frac{t \times 1 - (-y) \times 0}{t^2} + \frac{t \times 0 - (-y).1}{t^2}$$

$$a_x = \frac{x}{t^2} - \frac{x}{t^2} = 0$$

and y component of acceleration is given by

$$a_y = u\frac{du}{dx} + v\frac{dv}{dy} + \frac{dv}{dt}$$

$$a_y = \frac{x}{t}\times 0 + \left(\frac{-y}{t}\right)\times\frac{t\times 1-(-y)\times 0}{t^2} + \frac{t\times 0-(-y).1}{t^2}$$

$$a_y = \frac{y\times t}{t^3} + \frac{y}{t^2}$$

$$a_y = \frac{y}{t^2} + \frac{y}{t^2}$$

$$a_y = 2\frac{y}{t^2}j$$

Hence, answer is (D).

Q22. The given velocity field is

(A) Incompressible and rotational (B) Compressible and rotational

(C) Incompressible and irrotational (D) Compressible and irrotational

Solution:

For the flow to be incompressible, it must satisfy the continuity equation.

$$\frac{du}{dx} + \frac{dv}{dy} = 0$$

We have,

$$\frac{d}{dx}\left(\frac{x}{t}\right) + \frac{d}{dy}\left(\frac{-y}{t}\right) = \frac{1}{t} - \frac{1}{t} = 0$$

∴ Flow is incompressible

Now, we have

$$w_2 = \frac{1}{2}\left(\frac{\partial v}{\partial x} - \frac{\partial u}{\partial y}\right)$$

$$= \frac{1}{2}(0-0)$$

$$= 0$$

∴ It is irrotational flow

Hence, answer is (C).

Linked Answer Questions

Statement for Linked Answer Questions 23 and 24:

An incompressible fluid is passed through a T-junction supported on wheels, as shown in the figure. The area at outlet A is twice that of outlet B. While the incoming mass flow rate is fixed, the distribution of flow at the two outlets can be varied by a suitable mechanism built in the system. Assume that the flexible tube offers no resistance to motion, and frictional effects in the pipes and wheels can be neglected. Now, consider the following two cases:

Case 1: The flow rates at sections A and B are equal.

Case 2: The velocities at sections A and B are equal.

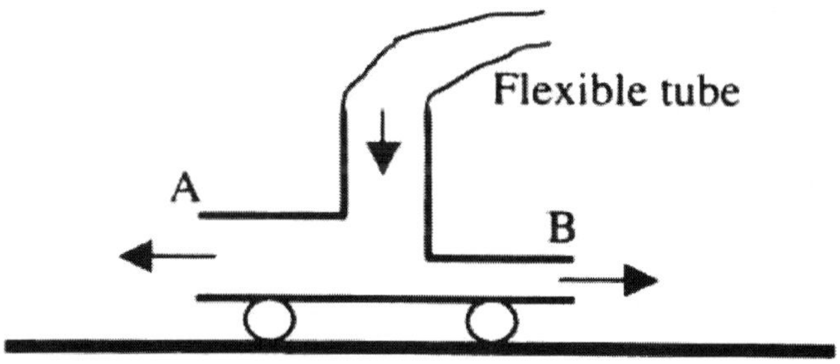

Q23. Which of the following statements are true?

P: In Case 1, the velocity at section A is twice the velocity at section B

Q: In Case 1, the velocity at section A is half the velocity at section B

R: In Case 2, the flow rate at section A is twice that at section B

S: In Case 2, the flow rate at section A is half that at section B

(A) P, R (B) P, S (C) Q, R (D) Q, S

Solution:

Given:

The area at outlet A is twice that of outlet B

i.e. $A_A = 2A_B$

Incoming flow ratio is fixed.

Case 1:

The flow ratio at section A and B are equal

i.e. $Q_A = Q_B$

$\therefore A_A V_A = A_B V_B$

$$\frac{A_A}{A_B} = \frac{V_B}{V_A}$$

$$2 = \frac{V_B}{V_A}$$

$V_B = 2V_A$ or $V_A = 1/2V_B$

i.e. Velocity of section A is half the velocity at section B

Case 2:

The velocities at section A and B are equal

i.e. $V_A = V_B$

$A_AV_A = A_BV_B$

$A_AV_A = 2A_BV_B$

$Q_A = 2Q_B$

i.e. Flow rate is twice that of section B

Hence, answer is (C).

Q24. Which of the following statements are true?

P: In Case 1, the system moves to the left

Q: In Case 1, the system moves to the right

R: In Case 2, the system moves to the left

S: In Case 2, the system moves to the right

(A) P, R (B) P, S (C) Q, R (D) Q, S

Solution:

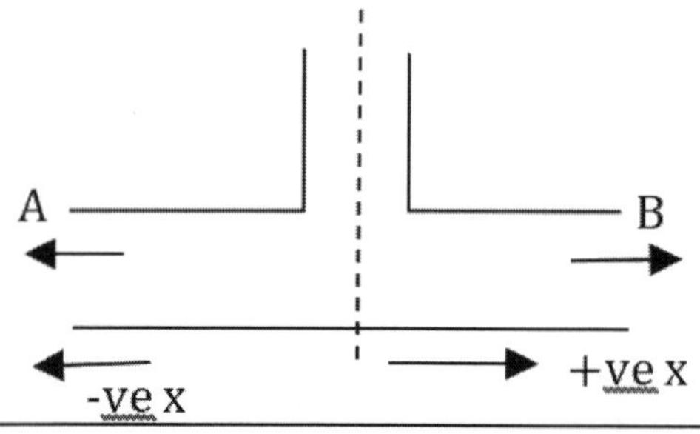

Note: Reaction force acts in direction opposite to the direction of the flow

Case 1:

The flow rate at section A and B are equal

i.e. $Q_A = Q_B$

Now reaction force at outlet is given by

$F_A = \rho\, Q_A\, V_A$ (in +ve x direction)

And reaction force at outlet B is given by

$F_B = \rho Q_B V_B$ (in -ve x direction)

Now, we have $F_A = \rho Q_A\; V_A = \rho Q_A \dfrac{Q_A}{A_A} = \dfrac{\rho Q_A^2}{A_A} \rightarrow$

And $F_A = \rho Q_B = \rho Q_B \dfrac{Q_B}{A_B} \rightarrow$ ...(ii)

$$\therefore \frac{F_A}{F_B} = \frac{\rho Q_A \cdot Q_A}{A_A} \bigg/ \frac{\rho Q_B \cdot Q_B}{A_B}$$

$$= \frac{\rho Q_A \cdot Q_A}{A_A} \times \frac{A_B}{\rho Q_B \cdot Q_B}$$

$$= \frac{Q^2}{Q^2} \times \frac{1}{2} \qquad (\because A_A = 2A_B \;\&\; Q_A = Q_B = Q)$$

$$= \frac{1}{2}$$

$$\frac{F_A}{F_B} = \frac{1}{2}$$

$$F_B = 2F_A$$

$\because$ F_B which is acting in negative x direction (left) is twice the F_A, which is acting in positive x direction (right)

$\therefore$ System will move in left.

Case 2:

The velocity at section A and B are equal

i.e. $V_A = V_B$

Now, we have $F_A = \rho Q_A\, V_A = \rho . A_A . V_A . V_A = \rho . A_A . V_A^2$

And $F_B = \rho Q_B\, V_B = \rho . A_B . V_B . V_B = \rho . A_B . V_B^2$

$$\therefore \frac{F_A}{F_B} = \frac{\rho A_A V_A^2}{\rho A_B V_B^2} = 2 \times 1 = 2$$

$\therefore$ $F_A = 2F_B$

$\therefore$ System will move in right direction

Hence, answer is (C).